MAKING KASHMIR BORDERLESS

The Regional Centre for Strategic Studies (RCSS) is a South Asian regional think tank, based in Colombo, Sri Lanka. Established in 1993, it is an independent, non-profit and non-governmental organization which encourages research, dialogue, and deliberation on a broad range of conventional and non-conventional sources of conflict. The RCSS enables scholars and other professionals to address, individually and collectively, problems and issues of topical interest for all South Asian countries.

The Centre's key objectives are to:

- Sponsor, coordinate, and support research on South Asian strategic, security, and environmental issues.
- Promote interaction among scholars and professionals in the region and beyond who are engaged in South Asian studies.
- Foster linkages and collaboration among institutions focusing on studies and activities related to conflict, conflict transformation, and regional cooperation in South Asia.
- Encourage a new generation of analysts and commentators to generate fresh ideas and perspectives on the security discourse in the region.

This publication was made possible by the Kodikara Award for South Asian Strategic Studies.

Contact address:

Regional Centre for Strategic Studies
68/1, Sarasavi Lane
Colombo-08
Sri Lanka
Tel: + (94) 11-2690913, 2690914
Fax: + (94) 11-2690769
E-mail: info@rcss.org
Website: www.rcss.org

RCSS Policy Studies 51

Making Kashmir Borderless

DEBIDATTA AUROBINDA MAHAPATRA

REGIONAL CENTRE FOR STRATEGIC STUDIES
COLOMBO

MANOHAR
2013

Published by
Ajay Kumar Jain *for*
Manohar Publishers & Distributors
4753/23 Ansari Road, Daryaganj
New Delhi 110 002
and
Regional Centre for Strategic Studies
68/1, Sarasavi Lane
Colombo-08
SRI LANKA
Tel: + (94) 11-2690913, 2690914; Fax: + (94) 11-2690769
E-mail: info@rcss.org
Website: www.rcss.org

First published 2013

ISBN 978-81-7304-989-7

Printed at
Salasar Imaging Systems
Delhi 110 035

Contents

Acknowledgements

I am thankful to Prof. R.A. Ariyaratne for his valuable suggestions on an earlier draft of the monograph and to Prof. Aparajita Biswas and Prof. P.L. Dash for their encouragement. I am also thankful to RCSS for supporting this study.

D.A. Mahapatra

Introduction

Kashmir has often been referred to as a 'dangerous place on earth' in various quarters, but seldom has its potential to evolve into a zone of peace been emphasized. The discourse has largely been state-centric and this study aims at widening the discourse by centrally locating the people of the region in the wider gamut of conflict and peace in South Asia. Kashmir, currently divided among three nuclear weapon States—India, Pakistan and China—remains an issue in conflict discourse, due to contested claims, militancy and terrorism. The cost and consequences of the conflict has become traumatic for the people in the region. The focus of this study is the contested and artificial division of Kashmir that has brought untold suffering to its people. The redrawing of boundaries has stalled the interaction among the people of Kashmir for nearly six decades and hampered economic development in a region that once was a trade hub in the Indian subcontinent spreading across Central Asia, China and beyond.

India and Pakistan, the States among which the major part of erstwhile princely state of Kashmir is divided, continues to address the issue from a narrow perspective though recently attempts towards an inclusive approach for conflict resolution have emerged. Despite border disintegration in various parts of the world, the state-centric approach to international relations in terms of upholding the sacrosanct nature of borders and absolute sovereignty still appears intact in Kashmir. The Indo-Pak conflict has led to repeated division of the region, a significant part has remained with India, and the rest divided between Pakistan and China, leading to the imposition of borders in an erstwhile undivided territory. The conflict over Kashmir between India and Pakistan started in 1947, when after Independence, both countries staked claim to the territory. Since then, it has taken a protracted turn resulting in three full-fledged wars, in 1947-8, 1965, 1971 and a limited war in 1999. The ceasefires that followed led to repeated drawing or adjustment of the Line of Control (LoC) in Kashmir, thus orchestrating recurring changes in its geographical contours.

These contentious developments have taken place at the cost of the people since the late 1940s. Borders were drawn and redrawn altering people's identity and space. The perpetrators remain in focus while those affected caught in the maze of hostile and contested divisions, remain largely neglected. This study has engaged with the people-centric approach of the contested borders in Kashmir and questioned the state-centric approach which is based on traditional notions of borders, security and territorial integrity. By positioning the people at the centre of the analysis, it seeks to rescue the geographic space that has tormented the people in this contested landscape. This volume intends to contribute to the emerging literature on border studies by focusing on the multiple dynamics of one of the most violent nuclear armed borders in the world.

NEW MEANING OF BORDERS

In the emerging global world order, borders have been transformed from rigid lines of separation to flexible lines of contact, commerce and co-existence. Borders worldwide are softening or even falling, as in Berlin. This realization needs to be brought to the conflict-torn region of Kashmir. Conflict in Kashmir can be transcended by exploring the potentials in terms of historical linkages and ties and economic complementarities in different parts of the region. One of the major steps in this direction will be the opening of all the traditional routes in the region. The opening of the Srinagar-Muzaffarabad route in 2005 and the Poonch-Rawalakote route in 2006 between the Indian Jammu & Kashmir (hereafter J&K) and the Pakistani 'Azad' Jammu & Kashmir (hereafter AJK),[1] had, to

[1] Pakistan has further divided the part of Kashmir under its control into AJK and Northern Areas (called Gilgit-Baltistan before the division of Kashmir).

some extent, accorded fluidity to the contested borders. Though travel and trade have commenced through these two routes, they remain largely symbolic, owing to various factors, including official apathy and, more crucially, the oscillating Indo-Pak relations in the aftermath of the 2008 Mumbai attack.

The historic initiatives in terms of opening cross-LoC routes need to be fostered and supported by India and Pakistan. This research, while documenting the positive impact of the existing cross-LoC contact, emphasizes the need to move beyond symbols and open all the traditional routes. The socio-cultural as well as economic significance and political ramifications of this move will be immense. Thousands of divided families (separated due to abrupt divisions created by the partition of Kashmir) will be able to meet each other. The Partition has also led to internal displacement of people who are eager to return to their native place or visit their families. Traders will benefit from the new markets in India and Pakistan. Trade benefits can go even beyond India and Pakistan if the Silk Route can be restored by opening the intra-Kashmir routes that once connected diverse parts of Asia and beyond. With immense tourism potential, a borderless Kashmir will also bring in prosperity. This crucial step will have wider implications for peace and security in the region. It will impact the political nature of the conflict and aid the conflict resolution process. The capital generated by the opening of the routes may, thus, help transform the region once pronounced the most 'dangerous place on earth' to one of the most peaceful and prosperous places.

This study assumes immense significance in view of the generation of peace capital due to the opening of the two intra-Kashmir routes and increasing realization on the part of the conflicting parties the importance of promoting peace and development in the region. It will map the costs of the rigid intra-Kashmir division and then document the positive impact of cross-LoC contact to make a case for a borderless Kashmir. By bringing to the forefront the socio-cultural and economic impact of the artificial division on the conflicting parties, this author intends to emphasize the urgency to move beyond symbolism and make Kashmir borderless by opening all routes to facilitate unrestricted movement of men and material across the region. Besides adding to the literature on conflict and peace, the research will have significant policy implications as it will provide an insight into the positive impact of the opening of the traditional intra-Kashmir routes to the region as well as the wider subcontinent.

NOTE ON METHODOLOGY

This study is the product of the review of available literatures on the Kashmir conflict, analyses of the news reports and extensive surveys conducted in J&K during the author's stint at the University of Jammu from 2005 to 2007 and later in 2008 and 2011. In all, the surveys through informal interactions and observation method, an attempt was made to ascertain the views of the people, especially those living in the border areas. In some instances, respondents were chosen for unstructured interviews through random sampling. The Akhnoor sector of Jammu region is worth mentioning in this context, where the author visited Chack Malal, Chack Rama, Rakh Malal, Jogwan, Sarmala, Kamdini Nallah, Mattoo, Dhaleri and Bhopur. In Poonch district, surveys were conducted in border villages near Surankote. The field visit also included Potha Billa and Khari Karmara, the latter of which is located on the Poonch-Rawalakote bus route near LoC. In Kargil, the author also visited border villages located amid difficult terrains such as Hunderman, Hunderman Brok, Kirkit Majdass, Kirkit Badgam, Kirkit Haral, Latoo and Kaksar.[2]

No single study is adequate to analyse the plight of people caught in the conflict. Besides, the tragedy that occurs due to the artificial lines of separation pervades people across the national, ethnic and religious boundaries. Another limitation of the study is that it does not cover the travails of the borderlanders living on both sides of the LoC throughout undivided Kashmir since the author could

[2] None of these villages are accessible to non-locals due to their proximity to the contested border. This author took special permission from the Senior Superintendent of Police in Kargil to visit these sensitive areas. This survey was part of my research as a Scholar of Peace Fellow with WISCOMP, New Delhi, 2007.

not personally visit AJK and Gilgit-Baltistan. But attempts have been made to ascertain their viewpoint. In this context, meeting people who travelled on the cross-LoC bus was helpful. Informal interactions with academicians, journalists and common people who came to J&K from the Pakistani side of Kashmir via the two opened bus routes and email interviews with people living across the divide have brought into focus the fact that the problems of the people, whether in AJK or Gilgit-Baltistan or J&K, are strikingly similar as they once shared a common socio-cultural identity. The urge to live in a borderless Kashmir is all-pervasive among the people of the region.

CHAPTER 1

Borders and Security

The collapse of the Berlin Wall and the subsequent end of the Cold War has led to the realization that the importance of borders is no longer effective in separating states. In fact, the border discourse has been widened to include areas revolving around the socio-cultural and economic aspects to make physical separation inconsequential. Most borders are no longer static, rigid and inflexible lines of separation. In this era of globalization where the trend towards mingling, multidimensional interactions and softening has gained currency, borders too have become dynamic. In the case of Kashmir, which enjoyed an integrated life for centuries, the border discourse has overtly emphasized the territorial dispute between India and Pakistan. While Kashmir's international image as the 'most dangerous place on earth' has been repeatedly highlighted, the contested border gets only occasional focus. India and Pakistan have fought four wars since the past 60 odd years, which directly or indirectly revolved around Kashmir. These wars drew and redrew borders and the militancy that surfaced in the late 1980s made the artificial divisions further sacrosanct and rigid.

This chapter explores the theoretical aspects of security and border, and its implications for Kashmir. It argues that the emerging discipline of border studies, which has impacted the traditional notions of border and security, has found in Kashmir an ideal case for exploring the intricacies of the contested border, making it flexible and humane. As the aim of this study is not to explore the history of Kashmir in detail, the chapter has provided an overview of the historical developments that initially led to the unification of distinct regions as a single princely state of Kashmir and later to its division. The transfer of power from the Sikhs to the Dogras, the further extension of the Dogra rule to the north, the subsequent British interest in the region due its strategic location and the fear of Russian intrusion, the partition of British India, and subsequent wars between the two independent nations have all shaped the history of the state in terms of its unification and division.

THE EVOLVING DISCOURSE

Since the emergence of modern states, borders have been an essential component of their making. The sacrosanct nature of borders has been one of the dominant themes of conventional studies as

> boundaries are viewed as constituting a given territorial fact, a static, unchanging feature, rather than which has its own internal dynamics and which influences, and is influenced by the patterns of social, economic and political development which take place in the surrounding landscapes—the frontier regions and/or borderlands.[1]

A traditional definition of state security implied defending the territory and keeping borders as signs and symbols of the sovereignty and territorial integrity of the state inviolable primarily through military power. Borders were perceived as 'markers of the limits of "national" . . . to be defended . . . not to be negotiable or flexible', and were defined as 'domains of contested power, in which local, national, and international groups negotiate relations of subordination and control'.[2]

In International Relations theory, this narrow definition of border is attributed to realism, one among the many contending theories that are at the core of the discipline. Realists consider

[1] Nurit Kilot and David Newman, 'Introduction', in Nurit Kilot and David Newman, eds., *Geopolitics at the End of the Twentieth Century: The Changing World Political Map*, London: Frank Cass, 2000, p. 9.

[2] Thomas M. Wilson and Hastings Donnan, 'Nation, State and Identity at International Borders', in Thomas M. Wilson and Hastings Donnan, ed., *Border Identities: Nation and State at International Frontiers*, Cambridge: Cambridge University Press, 1998, p. 10.

States to be at the core of international relations.[3] They view international politics as being competitive and conflictual wherein the relations among the States are governed by the Darwinian theory of survival of the fittest and the 'psychology of egoism'.[4] The actions of the States are governed by the narrow definition of security for gaining power, survival and pursuit of national interests in an international system that continues to remain anarchic and hierarchical.[5] Hence, a possibility of war among States, due to continuous struggle for power, can never be kept aside and, thereby, States are continuously engaged in developing military capabilities, both offensive and defensive, to protect the sanctity of their borders or even to violate the sanctity of the borders of the perceived rival State.

Of late, the retreat of realism has become part of international discourse with the approach being challenged by the emerging scenario.[6] Human security, as a challenge to ideas of traditional security, attempts to redress the asymmetry of attention to States at the cost of people in security discourse.[7] Here, the focus is on freedom from fear, danger and threat and the core referent for security is the individual, i.e. security should be people-centred and not State-centred. It is emphasized that the focus on people and human collectives is essential for making the concept of security inclusive and relevant to existing realities.[8]

[3] J. M. Grieco, 'Realist International Theory and the Study of World Politics', in M.W. Doyle and G.J. Ikenberry, eds., *New Thinking in International Relations Theory*, Boulder: Westview Press, 1997, p. 163.

[4] For details see D. Boucher, *Political Theories of International Relations: From Thucydides to the Present*, Oxford: Oxford University Press, 1998.

[5] Kenneth Waltz, 'Structural Realism after the Cold War', *International Security*, vol. 25, no. 1, 2000, pp. 5-41.

[6] M.D. Yaffe, 'Realism in Retreat? The New World Order and the Return of the Individual to International Relations Studies', *Perspectives on Political Science*, vol. 23, no. 2, 1994, pp. 82-3.

[7] Edward Newman, 'Human Security and Constructivism', *International Studies Perspectives*, vol. 2, no. 2, 2001, p. 240.

[8] For a detailed study of this approach see Barry Buzan, *People, States and Fear: An Agenda for International Security Studies in the Post-Cold War Era*, New York: Harvester Wheatsheaf, 1992.

The challenges posed to traditional definitions of security have significantly impacted the border discourse that too remained State-centric until quite recently. Borders have created many modern States with least attention paid to the identity, group relationships and shared culture of the people inhabiting the areas, thus giving rise to many inter-State border conflicts. The border can be a dividing line where on the one hand, the tie between citizenship and nationality has been broken, and, on the other, nation and state have been decoupled.[9] With the recurrence of inter-State border conflicts, the brunt is borne by citizens of countries on both sides of the border, but more prominently and continuously by borderlanders living in the contested landscape. The friction between rigid border and socio-cultural identity of borderlanders has, till recently, received sparse attention from policymakers as well as opinion makers and analysts.

The concept of security and consequently borders have undergone a sea change in the post-Cold War era that can be largely attributed to the era of globalization that puts forward internationalization and integration beyond borders. Free flow of money, men and material for growth and development seems to be the motto of this era. Hence, an argument has currently gained momentum in favour for humane borders. The recent border studies approach holds a crucial position in interlinking multiple disciplines such as International Politics, Political Science, Anthropology and Sociology to untie the intricacy of interplay of States and people in the contested zones. They have brought into focus the intricacies of border making and remaking and the interface of border with humanity by going beyond the traditional definition of border as a clearly defined and demarcated line between sovereign countries, forbidden to be crossed without the consent of the concerned state. To quote Kilot and Newman,

> The significance of just what boundaries are and what their impact on the ordering of political society is has also changed. From an almost exclusive focus on the physical nature and demarcation of the territorial lines, the

[9] Gerard Delanty, 'Irish Political Community in Transition', *The Irish Review*, no. 33, 2005, p. 13.

study of boundaries has become multidimensional, focusing on different scales of boundaries, as well as the impact on group and national identities, not all of which necessarily organized along territorial lines.[10]

The borders and border analyses are, thus, no longer confined to the realist paradigm of inter-State relations. It has gained increasing significance and wider relevance than a mere dividing line between States. They have entered wider discourse wherein the people-centric approach has emerged as a crucial component of the concept.

In the near future, a stateless international system is unimaginable. States are likely to continue to challenge each other for gaining power and ensuring their security. Security, in the narrow sense of the term, still continues to be a primary concern of States.[11] In the existing international scenario, despite the realization that borders can be made flexible, it is difficult for States to adopt a new definition of security wherein people are at the core and State security at the periphery. Thus, the first step towards transcending this traditional notion of security is to widen its existing definition. State and people-centric security cannot be mutually exclusive as they are interrelated and interlinked. As Edward Newman argues, the notion of human security does not 'exclude the importance of traditional ideas of security, but it does suggest that it may be more effective to reorient the provision of security around people—wherever the threat comes from. Traditional conceptions of State security—based on the military defence of territory—are an important but not a sufficient condition of human welfare.'[12] The security of people needs to be given as much importance as that of territory. But as of now, this too seems improbable in many parts of the globe including South Asia.

[10] Kilot and Newman, 'Introduction', op. cit., p. 13.

[11] R.J. Stupak and P. M. Leitner, 'Realism Revisited: Philosophical Assumptions, Power Patterns, and American Foreign Policy', *Journal of Power and Ethics*, vol. 2, no. 1, 2001, p. 86.

[12] Edward Newman, 'Refugees, International Security, and Human Vulnerability: Introduction and Survey', in Edward Newman and Joanne van Selm, eds., *Refugee and Forced Displacement: International Security, Human Vulnerability and the State*, Tokyo: United Nations University Press, 2003, p. 8.

In Kashmir, the realistic paradigm of inter-State relations in terms of sacrosanct borders, as markers of sovereignty and Independence of India and Pakistan, remained largely intact even after more than six decades of artificial divisions. Hence, the haphazardly drawn border in Kashmir is a unique example in the border discourse. The all-pervasive security apparatus in terms of observation towers, strict border surveillance, landmines and electrified fencing is testament to the impregnable nature of this border. In recent times, however, changes are perceptible in this trouble-torn region. People have been factored as a crucial aspect of the overall peace process. As of now, the security of people is not at par with the security of territory as far as the security system of India and Pakistan is concerned, but the crucial achievement from peace perspective is that both countries have at least factored people in the overall dialogue process. Concrete attempts have been made by India and Pakistan that have paved the way for the feasibility of a borderless Kashmir. The changing international scenario has brought forth the need for the two countries to chart out a middle path to resolve the long-standing Kashmir issue that has been a drain on their exchequers. Talks and gestures in terms of making borders soft and irrelevant, however, symbolic they may be, on the part of Indian and Pakistani leaders is a testament to the fact that changing concept of security and borders has reached the conflict-ridden region of Kashmir. The forces of globalization have provided the fertile ground for Indian and Pakistan to move ahead in the context of Kashmir and resolve it amicably for a better national, regional and international future.

EMERGENCE OF KASHMIR AS A SINGLE ENTITY

The erstwhile princely state of Kashmir can be geographically divided into three parts: Kashmir, Jammu and Ladakh. Historically, Kashmir had been under various rulers belonging to different religions including Buddhism, Hinduism, Islam and Sikhism. It had become a centre of learning as well as trade due to the Silk Route connections with Central Asia and China. Its rich historical and cultural legacy is reflected in ancient Hindu shrines, and Buddhist monasteries. Islam too contributed to its legacy. The famous *Rajatarangini* written by Kalhana in the twelfth century portrayed beautifully the history of Kashmir by narrating the reigns of various rulers. The Muslim rule dawned in the region in fourteenth century and Islam became the dominant religion. In 1588, Kashmir was annexed by Emperor Akbar and it became part of the Mughal Empire. Later, Kashmir was controlled by Afghan rulers. Kashmir Valley passed from the control of the Durrani Empire of Afghanistan to the Sikhs in the early nineteenth century. Till 1846, Kashmir remained a part of the Sikh empire.

The Jammu region to the south of Kashmir has its own historical and cultural legacy. According to a legend, the kingdom of Jammu was founded by Raja Jamboolochan in the fourteenth century BC. The name Jammu was derived from his name. The region was mostly dominated by Hindu Dogras. In 1780, the kingdom of Jammu was taken over by the Sikhs under Maharaja Ranjit Singh of Punjab. Till 1846, Jammu enjoyed the status of a tributary to the Sikh power until Gulab Singh of Jammu rose to prominence after his successful ventures in Gilgit and Baltistan by leading the army of Sikhs in 1819. For his great service to the empire, he was conferred the title of Raja (king) of Jammu in 1820. The Ladakh region, home to some of the world's highest mountain ranges such as the Karakoram, remained independent for nearly nine hundred years from the middle of the tenth century. The region was one of the most popular trade routes linking the Indian subcontinent, Central Asia, China and West Asia. Buddhism came to the region around AD 1080, when Sri Badat of the Shah Rais dynasty ruled it. After Sri Badat, who was the last Buddhist ruler of Gilgit, the Hunza ruler Shamsher founded the Tarakhan dynasty in Gilgit. Shamsher patronized Islam in the region. The reign of the Tarakhan dynasty was succeeded by Taj Mughal of Badakhshan in 1335. The region continued to witness invasions by Afghan rulers. Following the end of the Mughal rule in Kashmir during the nineteenth century, the region came under the rule of the Sikh king Maharaja Ranjit Singh in 1819. His reign came to an end when the British East India Company defeated the army of the then king Duleep Singh and signed the Treaty of Amritsar on 9 March 1846.

As per the treaty, the region was sold to Gulab Singh for Rs. 7.5 million. Article 1 of the treaty read,

> The British Government transfers and makes over for ever in independent possession to Maharajah Gulab Singh and the heirs male of his body all the hilly or mountainous country with its dependencies situated to the eastward of the River Indus and the westward of the River Ravi including Chamba and excluding Lahul, being part of the territories ceded to the British Government by the Lahore State according to the provisions of Article IV of the Treaty of Lahore, dated 9th March, 1846.

In return, the Maharaja, as mentioned in Article 3, promised to pay the British government the sum of Rs. 7.5 million out of which Rs. 5 million was to be paid on or before 1 October 1846. The British promised to provide aid to Maharaja Gulab Singh in protecting his territories from external enemies. Article 10 of the treaty read,

> Maharajah Gulab Singh acknowledges the supremacy of the British Government and will in token of such supremacy present annually to the British Government one horse, twelve shawl goats of approved breed (six male and six female) and three pairs of Cashmere shawls.

Gulab Singh gradually expanded his dominion to the remote Himalayan regions of Ladakh and Baltistan. The Dogras subdivided the region into two *wazarats* (an Urdu word which literally means ministry, in this case implying divisions), placing Hunza, Nagar, Ashkoman and some tribal areas under the Gilgit *wazarat*, while Ladakh and Baltistan came under the Ladakh *wazarat*. The region emerged as a single entity following the Treaty of Amritsar. Gulab Singh died in 1857 and was succeeded by Ranbir Singh (1857-85). Two other Maharajas, Pratap Singh and Hari Singh ruled the state in succession.

The emergence of the Soviet Union, after the Socialist Revolution of 1917, had caused concerns in the mind of the British. Besides, the cool climate of the region as also its strategic position attracted the attention of the British government. To quote E.F. Knight,

> The climate of this paradise of Asia appears to be well adapted to the European constitution. . . . Had we not sold this beautiful valley to the

Maharaja, a great military cantonment would no doubt have long since established here. This would not only have been most advantageous from a strategic point of view but would have avoided much of the sickness and mortality which thin the ranks of our white army in India.[13]

Similarly, Lord Curzon argued,

It is one of the northern gates of India, through which a would-be invader must advance, if he advances at all. Gilgit occupies a strategic place and the India government, harassed by Russia's growing restlessness in Central Asia, knew it for the key of great northern gateway into India, a key worth holding even at some cost in toil, money and valuable though less-human lives.[14]

The British set up the Gilgit Agency in 1877 with the appointment of Sir John Biddulph as the Political Agent. Hunza and Nagar were subjugated under the British Agency in 1891. In 1901, the Gilgit and Ladakh *wazarats* were reorganized and put under the control of the state government, while the remaining districts of the Gilgit Agency were put under the control of the Political Agent. In 1935, the British signed a 60-year lease with Maharaja Hari Singh under the Treaty of Jammu for what is now called the Northern Areas in Pakistan. The treaty stipulated that the state flag would continue to fly over the leased territory and Baltistan and Astore would remain under the Maharaja's control. The British recognized Maharaja's sovereignty over the whole region and as an acknowledgement of this factor the British garrison in Gilgit presented arms to the Maharaja on his birthday and other state occasions. The Gilgit Agency was handed over to the Dogra regime on 1 August 1947. Brigadier Ghansara Singh, along with the Chief of General Staff of the state army, General Scott and Captain Saeed Durrani arrived in Gilgit on 30 July 1947 and took over charge from Lt. Col. Beacon, the English Political Agent. The Maharaja then appointed Brigadier Ghansara Singh as his governor in the region. Hence,

[13] Quoted in N.N. Raina, *Kashmir Politics and Imperialist Manoeuvres: 1846-1980*, New Delhi: Patriot Publishers, 1988, p. 11.

[14] Quoted in F.M. Hassnain, *British Policy Towards Kashmir: 1846-1921*, New Delhi: Sterling Publishers, 1974, p. 61.

Gilgit-Baltistan became part of the princely state before India and Pakistan came into being on 15 and 14 August 1947 respectively.[15]

DIVISIONS AND EMERGENCE OF LINES OF SEPARATION

The princely state of Kashmir, the diverse and beautiful northern part of the Indian subcontinent, with sobriquets like 'Switzerland of the East' and 'paradise on earth', became a source of conflict between India and Pakistan in 1947. While the ruler of the princely state, Hari Singh, was toying with the idea of remaining independent, both India and Pakistan staked claim over Kashmir. The ruler requested a standstill agreement with both States that was accepted by Pakistan while India asked for further negotiations. Much remains disputed about the events that followed. According to one line of argument, Pakistan, which under the agreement was entitled to take care of supply and communication, stopped regular supply to the State as a pressure tactic. The uprising in Poonch, and later infiltration of armed groups from Pakistan in undivided Kashmir, complicated the situation. The invaders occupied the Muzaffarabad region of undivided Kashmir on 22 October 1947 and then advanced towards Srinagar in the Kashmir Valley. Hari Singh signed the instrument of accession with India on 26 October 1947. The legality of accession was contested by Pakistan. India contended that it had legal rights to the entire pre-1947 Kashmir. On 27 October 1947, Indian troops were sent to Kashmir and a full-scale war between India and Pakistan began. On 1 January 1949, the ceasefire negotiated by the United Nations created the line of division in Kashmir based on factual positions of the security

[15] It needs to be emphasized that the undivided princely state of Kashmir was not a State in the Westphalian sense of the term. British India consisted of provinces and princely states, among which the former were directly governed by the British and the latter by the rulers under British suzerainty. The Mountbatten Act of 1947, which led to the Partition of British India into independent States of India and Pakistan, provided that the rulers of princely states have to join either of the independent States while taking into account the wishes of the people.

forces of both India and Pakistan. The ceasefire line was delineated on maps during the Karachi 'Agreement of 27 July 1949, formally known as the Agreement between Military Representatives of India and Pakistan Regarding the Establishment of a Ceasefire Line in the State of Jammu and Kashmir.' By 3 November 1949, with the help of United Nations military observers, the borders on the map were demarcated on the ground. These developments led to the drawing of the first artificial divide in Kashmir. It also laid the ground for further drawing of borders with negative consequences for the region.

The artificial division led to a major portion of the princely state remaining with India, but also leaving a considerable portion in the possession of Pakistan. The ceasefire line, later rechristened LoC, divided Kashmir into three parts: AJK and Gilgit-Baltistan on the Pakistan side and J&K on the Indian side. The total area of the Kashmir region is 2,22,236 sq km, including 1,01,387 sq km with India, 78,114 sq km under the control of Pakistan, and 42,735 sq km under Chinese control, 5,130 of which was later handed over by Pakistan. AJK has all the decorations of a modern State such as legislature, an executive led by the president and the prime minister and a judiciary but is not recognized by the United Nations or any State as an independent entity. The other part, comprising Gilgit and Baltistan of undivided Kashmir, is called the Northern Areas. Initially, Pakistan placed it under the administration of AJK, but later took over its administration.[16]

The division of Kashmir in the late 1940s did not settle the issue. In the following decade, the borders were redrawn in the aftermath of the 1965 and 1971 wars. Most of the later divisions in Kashmir had taken place in the Ladakh region. Though most of the Kargil region had been with India since the first Indo-Pak War, the Kargil heights were under the control of Pakistan. The area was taken over by India in the war of 1965 but later, under the

[16] For details on AJK and Gilgit-Baltistan, see Debidatta Aurobinda Mahapatra and Seema Shekhawat, *Kashmir Across LOC*, New Delhi: Gyan Publishing House, 2008, pp. 115-42.

Tashkent Agreement, it was handed back to Pakistan. In the 1971 War, the Kargil heights were recaptured by India. Following the ceasefire, each side retained their respective gains. India retained the Kargil heights in line with the Simla Agreement of 1972. The agreement stipulated that the border would be delineated based on actual control of the region by the two armies on the day of ceasefire. In the case of Chhamb area in the Jammu region, which was taken by Pakistan during the 1965 War but recaptured by India in the 1971 War, it was returned to Pakistan after the war under the Simla Agreement. The contested border in Kashmir that was earlier known as the Ceasefire Line was rechristened the Line of Control (LoC), with some alterations based on actual possession of territory at the time of the ceasefire in December 1971. The 778 km LoC has been the *de facto* border in Kashmir since 1972. The redrawn borders in Kashmir following the 1971 war remained sacrosanct until the Pakistani intrusion in Kargil in 1999 but this limited war ended without any redrawing of borders.

Militancy started in the late 1980s in the Kashmir Valley inhabited predominantly by Sunni Muslims. In 1988, a series of violent demonstrations erupted on various issues such as the hike in power tariff and the demand for a ban on Salman Rushdie's *The Satanic Verses*. Pakistan's Independence Day was celebrated with fanfare on 14 August; whereas India's Independence Day on 15 August was marked by protests. The day of the accession of Kashmir to India, 26 October, was denounced as the day of occupation. The death anniversary of J&K Liberation Front leader Maqbool Butt was commemorated while Sheikh Abdullah's death anniversary was termed *Yom-e-Nijat* (Day of Deliverance). By 1989, a number of significant armed groups began to operate throughout the Valley to fight for *azadi* (freedom). Violence spread to the Muslim-majority hilly areas of Jammu including Poonch and Rajouri districts. Analysts have subscribed to different overlapping theories to explain the rise in militancy—ranging from unfulfilled Indian commitment of conceding right to self-determination to dilution of special powers accorded to J&K; from lack of democracy to patronage role of Pakistan. The tangible and intangible costs of violence due to militancy caused far reaching economic, cultural, social devastation in the region.[17] This also led to increased rigidity at the borders separating the two parts of Kashmir. The border became more violent with increased tension reflected in the number of security personnel, permanent laying of mines at all possible places to check cross-border infiltration by the militants, electrified fencing and observation towers.

[17] For a detailed study on the costs of Kashmir conflict see, Debidatta Aurobinda Mahapatra and Seema Shekhawat, 'The Peace Process and Prospects for Economic Reconstruction in Kashmir', *Peace & Conflict Review*, vol. 3, no. 1, 2008, pp. 1-17. Debidatta Aurobinda Mahapatra, 'Symbiosis of Peace and Development in Kashmir: An imperative for Conflict Transformation', *Conflict Trends*, no. 4, 2009, pp. 23-30; Also by the same author, 'Conflict and Development in Kashmir: Challenges and Opportunities', in Hari Dhungana and Marty Logan, ed., *Sustainable Development in Conflict Environments: Challenges and Opportunities*, Kathmandu: Centre for International Studies and Cooperation, 2007, pp. 68-78.

CHAPTER 2

Cost of Artificial Divisions

Border areas possess characteristics that differentiate them from the mainland. They determine relations between States, their origin and later developments. The impact of rigid and tense border on the people living around it is immense as it leads to specific costs for the people living in the contested spaces. Borders are established to separate or amalgamate States but when they rupture the socio-cultural identity of communities, the intensity of suffering increases manifold; as in the case of Kashmir, when links such as kinship, residence, tradition and interests transcend the physical limits of the state and the people residing across those borders share an integrated identity unrecognized by the state, there develops a friction between the rigid borders and the aspirations of the people. Pushed to the edge of power contestation between two powers, India and Pakistan in the case of Kashmir, the ordeals of the those living on sacrosanct borders is perilous.

No reliable data is available but Kashmir has suffered severe casualties in terms of men and material losses. This chapter provides an overview of the cost of the conflict, which largely accrues as a by-product of the relations between India and Pakistan and their contested policies on Kashmir, which mostly do not take into account popular aspirations and concerns. Hence, a tense border during times of bilateral hostilities increases the intensity of the problems, while one may witness relatively tranquil borders during times of dialogue.

RECURRING DISPLACEMENT

Border residents get displaced whenever the border is disturbed due to wars, war scares, heavy firing, shelling or even the mobilization of security forces. Though the borderlanders are not participants in such hostilities, they bear the cost in multiple ways including displacement. Border displacement is not always accidental but sometimes deliberately thrust upon people when they are directly targeted by cross-LoC firing and shelling. As Skjelsbaek and Smith argue, 'They (civilians) are no longer . . . part of the "collateral damage" consigned to the margins as perhaps regrettable and probably unintended, but unfortunately, inevitable casualties of military exigencies . . . in many wars, the civilians are the targets.'[1] The entire population of border villages have to abandon their homes and take refuge in interior parts when borders become extremely tense. There are different categories of displaced borders: some are permanently displaced while other groups shuttle between their homes and camps depending on the situation on the border. In case of Kashmir, the history of border and displacement coincide. In 1947-8, there was huge influx of about 50,000 Hindu and Sikh families from the areas of Muzaffarabad, Mirpur, Bagh, Rawalakote, Bhimber, Kotli and Jhanger in AJK to J&K. At present, over one million AJK displaced people are residing in various parts of J&K, mainly, Bhor, Digiana and Gole Gujral in the Jammu district. The displaced communities appear to have reconciled to the fragile terms of their new life and are more or less permanently settled in J&K. Initially, they lived in camps for many years with the hope of returning one day to their native places in AJK.

The 1965 war again led to large-scale displacement of borderlanders. During this war, around 1,00,000 people were forced to escape from Chhamb-Jaurian area in Jammu. The displaced were lodged in eight tented camps in the mainland of the Jammu province. Following the signing of the Tashkent Declaration between India and Pakistan in 1966, most of them returned to their homes. The 1971 war again led to displacement from the Chhamb area. Estimates suggest that 4,900 families were displaced from ten villages. Around 2,400 families were accommodated in tented camps at Kishanpur-Manwal in Udhampur, where they had to stay for

[1] Inger Skjelsbaek and Dan Smith, *Gender, Peace and Conflict*, New Delhi: Sage Publications, 2001, pp. 3-4.

about five years. Later, they were officially registered as displaced and settled in different parts of the Jammu region, especially in Jammu and Kathua districts on the border. The government of India announced relief packages in April and August 2000 for the 'non-camp displaced persons' from the Chhamb area and displaced persons from AJK. A committee headed by the Divisional Commissioner, Jammu, was appointed to verify whether the claims were genuine or not of the eligible displaced persons. The following benefits were provided to the displaced:

- Payment of ex-gratia at the rate of INR 25,000 per family to 'non-camp displaced persons' from Chhamb-Niabat area (1971);
- Payment of ex-gratia at the rate of INR 25,000 per family to the displaced persons from AJK (1947);
- Payment of cash compensation in lieu of land deficiency at the maximum rate of INR 25,000 per family of displaced persons from AJK (1947);
- Payment of INR 20 million to be provided for the allotment of plots to those displaced persons who have already been settled in J&K and who have not been allotted plots in the past;
- Payment of INR 2.5 million for improvement of civic amenities in 46 regularized colonies of displaced persons.

A total amount of INR 61.7 million was released by the Government of India to the Government of J&K for disbursement to eligible families. The Government of India has released a further amount of INR 490 million to J&K government on 24 December 2008 on account of payment of ex-gratia for land deficiency to displaced persons from AJK in 1947 following the package announced by the Indian Prime Minister Manmohan Singh in April 2008. Out of INR 490 million, an amount of INR 207.5 million has been disbursed to 2,004 eligible families/beneficiaries by J&K government. Regarding the 'non-camp displaced persons' from the Chhamb-Niabat Area (1971), the committee has verified 1,502 cases out of a total of 1,965 cases for payment of ex-gratia INR 25,000 per eligible family. The Government of India released INR 8.3 million to the J&K government in March 2004 for disbursement to the eligible beneficiaries. As many as 1,198 beneficiaries have been paid the specified compensation so far.[2]

The Kargil war in 1999 resulted in the displacement of a considerable number of people from the borders. The war scare of December 2001 after the attack on the Indian Parliament led to massive displacement from the border. Another war scare following a *fidayeen* attack in J&K on 14 May 2002 also led to displacement. As per an estimate, between the Kargil war and the 2002 war scare, over 1,75,000 people in the state were forced to leave their native places from borders—22,000 from Poonch, 9,000 from Rajouri, 1,25,000 from Jammu and 25,000 from Kathua district.[3] These developments indicate that displacement has become a part and parcel of the life of the borderlanders. They keep ferrying between their native place and camps as nomads since many of the residents from the border areas have been uprooted nearly seven times since the late 1940s (1947-8, 1965, 1971, 1987, 1999, 2001 and 2002). The period of displacement ranges from a few days to months and sometimes even years. Thus, the displacement is largely temporary but recurrent.

In a survey conducted in the camps set-up on the fringes of Jammu city to accommodate the borderlanders from the Akhnoor sector during the 2001-2 standoff between India and Pakistan, it became obvious that the situation in the camps is quite poor. The people are herded together without basic amenities such as proper shelter, water, or toilet facilities and live in impoverished conditions. Often they are at the mercy of public authorities for relief, which is meagre and irregular. As Punno Devi pointed out during the survey,

> We are living like animals herded together. These tents are like cages . . . no drinking water, no toilet and health, no privacy . . . we have lost our livelihood . . . transport is not available and we have to walk a long

[2] Ministry of Home Affairs, Government of India, *Annual Report 2010-2011*, pp. 249-50.

[3] For details see Seema Shekhawat, *Conflict and Displacement in Jammu and Kashmir: The Gender Dimension*, Jammu: Saksham Books, 2006, p. 98.

distance to reach roads to get public transport . . . education of our children has been disrupted . . . there is no socio-cultural life . . . actually there is no life . . . there is no dignity . . . we don't live, just run . . . from our homes to camp and then from camp to villages.[4]

Border residents are mainly engaged in agricultural activities. As per an estimate by a prominent local daily of 3 June 2002, nearly 3,50,000 hectares of agriculture land lay uncultivated due to tension-induced dislocation from the border areas.[5] This has led to heavy losses for the people. Besides, the non-cultivation of land for quite some time makes fertile land barren, thus increasing the burden of losses. During Operation Parakaram, due to heavy deployment of security forces and mining launched by the Government of India following the Parliament attack, people were unable to till their lands for two seasons. The de-mining process was not completed for long after that, and many farmers had to stay away from their lands.

There are cases of displaced people who have chosen to settle in the interior parts of J&K but most borderlanders have no option but to return to their homes as soon as the crisis abates. The Kargil war led to the displacement of thousands from all over the border including the Akhnoor sector in the Jammu region. A survey conducted in 2006 revealed that about 48,000 people had returned to their respective border villages but about 12,000 continued to live in three camps—Devipur, Naiwala and Rampur colony—even though the limited war had ended in 1999.[6] The survey revealed that many of the returnees did not return voluntarily. They were forced by the authorities through various tactics including closing of camp schools and health dispensaries, and stopping supply of drinking water, electricity, and distribution of relief. Ram Pal living in Naiwala camp lamented,

We return only to prepare ourselves for another displacement. . . . It is

[4] Ibid., p. 104.

[5] *Daily Excelsior*, 3 June 2002.

[6] Seema Shekhawat and Debidatta Aurobinda Mahapatra, *Kargil Displaced of Akhnoor in Jammu and Kashmir: Enduring Ordeal and Bleak Future*, Geneva: Internal Displacement Monitoring Centre, 2006.

a choice between death and devil. Life on the border is hell . . . here in camps it is no better . . . but at least we are alive here . . . in our native place no one knows when a person will get killed by firing or even mine.[7]

FIRING AND SHELLING

The borderlanders living within 5-7 km of the line of control have to face frequent firing and shelling from across the border. A small trigger, such as a belligerent statement from a political leader from either side, leads to firing and shelling. There are occasions when exchange of heavy firing including artillery and mortar takes place without any apparent reason. For this state-centric dynamism of the border the ordeal of the borderlanders appears largely a non-issue. No reliable data is available on the exact number of casualties due to firing and shelling on the borders, though it is commonly agreed that hundreds of people have lost their lives besides incurring material losses in terms of destruction of houses and other immovable property as well as livestock. The nature and extent of the devastation can be gauged from a local newspaper, which reported the killing of 72 people in cross-border firing between May 2001 and 2003.[8] According to another report, between January and 15 May 2003, firing took place as many as 1,007 times, claiming the lives of 29 people, cattle and damaging 49 houses.[9] Sometimes people are not able to harvest their ripe crops due to firing, fearing loss of lives. As per a report, in the Ranbir Singh Pura Sector of Jammu region, 800 to 900 acres of agricultural land came in the range of fire on 4 May 2002,[10] when India and Pakistan were engaged in coercive diplomacy on the border in the aftermath of the Indian Parliament attack. This resulted in the burning of ripe crops in the fields. There were instances of frequent clashes on the intra-Kashmir border, especially after the onset of militancy, before the nuclear-armed neighbours agreed to a ceasefire in 2003.

[7] Ibid.

[8] *Daily Excelsior*, 10 May 2003.

[9] *Daily Excelsior*, 22 May 2003.

[10] *Daily Excelsior*, 9 May 2002.

They now exchange sporadic fire. Over the past eight years, the instances of firing have reduced significantly.

FORTIFIED DEFENCE

Another ordeal faced by the border people is related to excessive measures displayed by the rival powers—India and Pakistan—in fortifying the lines of separation with a multilayered security structure comprising mining, Ditch-cum-Bandh (DCB), fencing and presence of defence personnel. During the time of actual Indo-Pak hostilities in 1965 and 1971, mines were planted all along the border, in cultivated land and pastures, around infrastructures and even houses, to obstruct movement from across the border. In the late 1980s, with the rise of militancy in J&K, heavy mining in border areas was undertaken purportedly to check cross-border infiltration, and to stop support and patronage from across the border. Besides, at the time of heightened tensions, for instance, during Operation Parakram, the armed forces mined most of the border in the last week of December 2002.[11] The Indian army took under its control a total of 70,100 acres of land in Jammu, Kathua, Rajouri and Poonch districts after the deployment of forces. As per unofficial estimates, over 25,000 acres of land was mined with Anti-Personal Mines (APMs) and Anti-Tank Mines (ATMs) with a density of 1,000 mines per square km. In Kathua and Jammu districts, the army reportedly took over 31,927 hectares of land of which 23,078 hectares became a 'literal minefield'.[12] Unlike the mining of 1965 and 1971, the mining of 2002 was widespread. While in the earlier cases only the radius of 1 km area was mined, in the latter case more than 3 km was mined.

The end of hostilities as witnessed in past cases normally leads to the removal of landmines from habited areas. But many mines remain undetected. Hence, deaths and injuries due to mine explosions are common. During the survey in the Kargil border villages, the author noticed landmines planted in 1965 and 1971 on

[11] *The Kashmir Times*, 13 January 2002.

[12] *The Kashmir Times*, 8 February 2004.

both sides of the narrow walkway about 6 ft wide on the mountain terrain. Though the visible surface of the mines looks faintly like a thrown away tin can, the local people cautioned that these mines are powerful enough to take a person's life or permanently incapacitate him. People remain cautious but occasional strays prove fatal. There are instances of people crippled by these mines while collecting firewood or cutting grass as they accidentally activate them. 'I was doing my normal activity of grass cutting in the morning. When I lifted my right foot to take the next step, there was a thundering sound almost making me deaf. Seconds later, I realized I had stepped on a landmine. I lost a leg', said Shabana in Jaurian area of Jammu. Domestic and wild animals also fall prey to these mines. The exact number of victims of landmines is not available, but an estimate of heavy physical losses can be gauged from the fact that in Chagia, a small village in Ranbir Singh Pura sector of Jammu, landmines set up during the 1971 War have injured as many as 23 residents till 2001.[13] As per another report, more than 2,000 victims of landmines had been recorded in the Rajouri-Poonch belt between 1947 and 1989.

In October 2004, the authorities claimed that de-mining operations have almost been complete. The author's survey in the villages of the Akhnoor sector in 2006, where the people returned, brought forth the fact that mines still remain undetected and, hence, are a continuous threat to the people. They not only cause physical injury, but their potential in terms of generating fear in the minds of people is phenomenal. In the Panjtoot village, about half a kilometre from the border, mines have been discovered in the fields leading to casualties in recent years. Though no human casualty was reported during the author's survey, there were cases of cattle being injured or killed due to undetected mines in January-February 2006. On conditions of anonymity, an army official admitted that it is not feasible to entirely de-mine the area since mines change their positions due to soil movements caused by rain or rodents. As India and Pakistan are not party to the Mine Ban Treaty that prohibits member countries the option of using,

[13] *The Kashmir Times*, 13 January 2002.

stockpiling, production and transfer of anti-personnel mines, there is a lack of binding laws which can give pressure to both the countries to discontinue the practice of mining.

Fencing and floodlighting are considered important for maintaining vigilance along the border. Barbed wire fencing has been erected by the Indian security forces to check infiltration from across the border.[14] The fence generally remains about 150 yards on the Indian side and consists of a double-row of fences and concertina wire ranging from 8 to 12 ft in height. It is electrified and connected to a network of motion sensors and alarms in sensitive parts to check violation. The stretch of land falling between the two rows of fencing is also mined. Fencing was started all along the J&K border in the 1990s. It slowed down in 2001-2 as hostilities increased resulting in exchange of fire at the border, thereby, disrupting the fencing work. After the November 2003 ceasefire agreement, fence building was resumed and fencing in Kashmir Valley and Jammu region was completed in 2004.[15] The fencing, done at a distance of 2-5 km instead of the edge of the contested border, has covered large areas of cultivable land amidst the actual agreed border and fencing area. This led to curtailing of the mobility of the people and affected access to lands. In many cases, the land remains uncultivable, as the landowners have to cover a distance of about 2-3 km to reach the gates made in between the fencing to access their land and that too during fixed hours which otherwise is located at a few metres away from their houses. The barbed wire fencing proves critically dangerous and life threatening at places where they are electrified, as livestock and stray animals often fall prey to it. Recently, to facilitate the border inhabitants to cultivate their lands without constraint, the government has decided to shift 38.015 km of fencing erected in the Jammu area,

[14] The available information reveals that the infiltration that consistently decreased since 2005 has reversed in the year 2009 and increased substantially compared to 2008. Ministry of Home Affairs, Government of India, *Annual Report, 2010-2011*, p. 6.

[15] 'LoC Fencing Completed: Mukherjee', *The Times of India*, 16 December 2004.

closer to the international border. But this is a small patch of land. The same approach needs to be extended all along the LoC. The fortification of the border is further aided by DCBs. *Bandhs* (raised structure made by piling huge amounts of soil) have been built on both sides as a barrier to prevent rival forces from watching the activities of their troops. The ditch provides a secure hiding place to the security forces during firing and shelling. These DCBs are, on many occasions, built on the land of the borderlanders, thus adversely affecting their livelihood. This usually inert multi-tiered defence mechanism becomes abuzz with the presence of defence personnel, the strength of which varies in accordance with the prevailing situation at the border. The strength may vary but their presence is permanent in and around the border villages, thereby restricting the mobility of borderlanders. At many places, the border people have to carry identity cards while travelling to and from their villages. Education, health services and other essential facilities like transport and communication too have taken a backseat in the border areas as the tense atmosphere discourages development activities. The relatives from mainland shy away from visiting them, lest any misfortune of border life befall them. Residents find it difficult to get suitable marital prospects for their sons since people from interior parts do not wish their married daughters to live on the tense border. Leading a peaceful life has become a distant dream for border residents.

There are provisions for compensation for those who die in the firing, for the houses destroyed and cattle killed, and for farmers whose lands have been covered under fencing or taken for erecting DCB or have been mined. The compensation is, however, meagre and the procedure is rife with malpractices. In many cases, even this meagre amount does not reach the victims in time due to widely prevalent malpractices. The promised compensation is either delayed or only partially paid or is paid in instalments or simply remains in indefinite process of reaching the victims. There are instances when victims did not receive any compensation on the facetious plea of non-verification or unavailability of proof of damages, further adding to the woes of the border people.

DIVISION OF FAMILIES

A relatively unknown dimension of the suffering of people is the forced division of families due to the sudden, haphazard and artificial creation of borders. When the borders were drawn and redrawn after the wars, men and women who happened to be on either side of the border, were forced to remain there without any recourse to return to their families and native places. The widest possible protection and assistance should be accorded to the family, which is the natural and fundamental group unit of society, reads Article 10(1) of the International Covenant on Economic, Social and Cultural Rights. Thousands of families living on either side of the LoC are denied this right following the partition of Kashmir. Families were split—women lost their husbands, mothers lost their children, and sisters and brothers were separated from each other. It is commonly agreed that thousands of such families are spread across the LoC, cutting through villages and houses, leaving behind a long history of suffering. The divided families find it difficult to come to terms with this arbitrary division and consequent impromptu separation. Kirkit Badgam, situated on a hilltop at about 5 km from the LoC, came on the Indian side after the 1971 War. Since then, 70-year-old Nazia, living in the village, has not been able to meet her husband who had gone for work to Skardu and got stranded there due to the ceasefire. Nazia only receives occasional letters from him with news of his remarriage, birth of his children and now their marriages. Eighty-seven year old Marriam has been waiting to meet her son whose whereabouts are unknown since the 1971 War but she believes that he is alive and lives in Skardu. In 1995, she even wrote an emotional letter to the President of Pakistan to trace her son and inform him that his father had died, but there has been no response. For years, Jammu-based Noor Mohammad, 86, had been nurturing a dream to meet his daughter who lives in Sialkot, just 11 km away from the border at Suchetgarh. He says, 'For us it will be a happy family reunion after decades of separation. I had written letters several times but there has been no response. Perhaps she is not getting them. There is a lot of news I have to give her including the death of her mother four years ago.' Hundreds of such stories are scattered throughout the border landscape of Kashmir.

Before the 1971 War, the border was comparatively porous. Crossing the border, though illegal, was not very difficult, thus enabling at least some families to remain in contact. There are narratives of people crossing the border by covering long distances on foot across difficult terrains. Even after six decades, 85-year-old Rubiya Begum of Hunderman village in Kargil, has not been able to reconcile with the tragedy of separation from her husband, who had gone to Skardu (now on the Pakistan side) for shopping and could not return due to the abrupt setting of the borders. With no news from her husband for two years, Rubiya tried to cross the divide in 1950 but was caught by the Pakistani security forces and handed over to the Indian forces. Her wish for reunion remains unfulfilled. After the onset of militancy, even such attempts came to a halt. The intricate linkage between the external dimension (Indo-Pak conflict) and internal dimension (militancy in J&K) added to the ordeal of these people. Initially, the LoC was a line of separation, but militancy changed its character—and made it a line of infiltration for the militants. The state paraphernalia became more rigid to check infiltration, thereby restricting communication. The artificial divide, hence, became inviolable.

People tried to remain in contact with their separated families but it was not an easy task. While for some getting the exact address was a problem, for others exchange of news is a lengthy process which includes waiting for months, as letters have to go through scrutiny by security forces on both sides, and then travel through a long and circuitous route. Letters cannot be despatched directly to the families and friends living at a distance of even less than 5 km across the LoC. In case of border villages in the Kargil sector, these letters have to pass through the Kargil head post office, New Delhi-Islamabad-Gilgit-Baltistan and then to the intended recipients. The detour takes two to three months and sometimes even six months. The news of Noor Mohammad's death in Skardu, reached his wife Hakima Begum in Kaksar in the Indian side after eight months. Hakima was in Kirkit Majdass to meet her parents but remained stranded after the ceasefire of 1949. She tearfully narrated her story.

It was a shock. I could not meet my husband in his life time and could not even mourn his death in time. Look at the tragedy; the day he died, we were celebrating the marriage of our grandson. After waiting for ten years, my husband remarried in Skardu but I did not. I raised my only child single handedly, married him and now even his son has got married. We were celebrating here and there my husband was put to eternal rest. Not even once our family could reunite. Now with my husband's death, there is no hope for reunification.

The post-1989 era witnessed stringent restrictions even on the limited communication facilities. With the advent of militancy in the Kashmir Valley, the lines of communication along the border were further cut off. No phone calls were allowed across the LoC. Militancy also affected emotional ties and raised the suspicion of the authorities on the possible involvement of the local people in terrorism. Many people refrain from keeping in touch with their relatives. Some do not initially acknowledge their cross-LoC connections, but after interactions, they confess their longing for reunion. With a passport in hand, people had to travel 1,700 km from Kargil to reach Skardu via New Delhi and Islamabad (which is actually less than 200 km from Kargil) to meet their relatives. Some villages across the LoC are situated at a distance of even less than 10 km but covering that distance is not easy. The complicated procedures of passport and visa acquisition, and economic constraints have ensured that reunion remains a dream for most. The divided families confront the challenges of life in numerous ways. Though many of them have moved on in life, the pangs of separation still overwhelm their hearts and minds.

CHAPTER 3

Soft Border: Humanitarian Implications

A sustained peace process in recent years with obvious ups and downs has brought normalcy to the border areas with significant implications for the people. But their problems have not ended with the ceasefire and other peace initiatives. However, the developments played an effective role in reducing the trust deficit between India and Pakistan and in humanizing the border at least partially towards ameliorating the suffering of the people. Since 2003, people have not been displaced, the frequency of firing and shelling has gone down, and more importantly, the perceptible decrease in violence on the border has brought respite for borderlanders and other stakeholders in peace, thus raising the hopes of a peaceful resolution of the Kashmir issue. These developments have led all stakeholders to think in terms of a flexible, friendly and pacifist border in one of the most troubled regions in South Asia.

Under Indo-Pak peace initiatives, the border was softened by opening intra-Kashmir routes to enable the meeting of divided families and later for trade. The gesture is quite encouraging and its dividends are notable in terms of making a rigid and sacrosanct border, soft and tractable, as a middle path between the options of non-existence of LoC and according it the status of an international border. Going beyond the traditional viewpoint of sovereignty and territorial integrity, under this approach, the LoC continues to remain a dividing line but allows movement of people and goods. However, the movement of men and material via the opened routes is limited and uneven. Also, the flexibility accorded to the LoC has remained confined to some regions, thereby only partially addressing the plight of the people. This chapter analyses various initiatives towards softening of borders in Kashmir, the humanitarian implications and limitations.

PEACE ATTEMPTS

Geopolitical changes brought about by the end of the Cold War altered the dynamics of international relations with implications for conflicts around the world. The rising profile of peace and conflict resolution constituency brought a new perspective to pacification and transition in the conflict-prone regions. Conventional armed disputes have been transformed by the intensification of globalization and increasing acceptance of peaceful means for conflict resolution.[1] The multiple players involved in the complex Kashmir conflict—India, Pakistan, Kashmiri people, and the international community—have been influenced by these developments. The main stakeholders, India and Pakistan, have adopted a multi-pronged strategy including initiation of dialogue between New Delhi and Islamabad as well as between New Delhi and Srinagar. Peace attempts between India and Pakistan in the emerging post-Cold War world order reflected the impact of the changing global environment in terms of softening of borders. Pakistan, softening its rigid position that it would not agree to a resolution based on thc *status quo* and India, softening its stance that a resolution would not involve compromise on boundaries was no mean achievement. This change of positions to chart out a middle path led to the crafting of peace measures with concrete results, among which a flexible intra-Kashmir border emerged as a prominent one.

The peace process gathered momentum with the accession of the National Democratic Alliance (NDA) to power in New Delhi in 1998. The NDA government initiated peace measures, notably with a historic bus journey from New Delhi to Lahore on 20 February 1999 with Indian Prime Minister Atal Behari Vajpayee on board. But with the Kargil episode, the peace process came to a

[1] Debidatta Aurobinda Mahapatra, 'A Perspective on Peace in Kashmir', *ICFAI Journal of Governance and Public Policy*, vol. 2, no. 4, October-December 2007, p. 31.

halt, though it could not permanently obstruct the movement for peace in the region. On 18 April 2003 Vajpayee extended a 'hand of friendship', which was grabbed by Pakistan.[2] In October 2003, India proposed a series of confidence-building measures aimed at improving communications by road, rail and sea between the two sides. On 26 November 2003, a ceasefire on the dividing line in Kashmir was announced. Since its creation in 1949, this area had remained tense experiencing shelling and firing even during periods of relative peace. For the first time, India and Pakistan agreed to completely halt hostilities on the border, including cessation of firing and shelling. Despite reports of occasional violations, the ongoing ceasefire is significant as the first formal ceasefire agreement between India and Pakistan since the outbreak of militancy in J&K. Its continuation to date has brought perceptible normalcy to border areas and its inhabitants, who have borne the brunt of the tumultuous border. In October 2009, during a visit to J&K, Indian Prime Minister Manmohan Singh acknowledged:

> We had the most fruitful and productive discussions ever with the government of Pakistan during the period 2004-07 when militancy and violence (in J&K) began to decline. Intensive discussions were held on all issues including on a permanent resolution of the issue of Jammu and Kashmir.[3]

This ceasefire, in effect till date, paved the way for many historic developments in terms of movement of people and goods across the hitherto sacrosanct dividing line.

PERMEABLE BORDER

The post-Cold War global changes, increasing constituency for peaceful resolution of conflicts and civil society movements in both parts of Kashmir as well as India and Pakistan, impacted the leaders of both countries. The leadership of both countries realized the importance of involving people in the peace process.

[2] 'PM Extends "Hand of Friendship" to Pakistan', *The Hindu*, 19 April 2003.

[3] 'Show Sincerity in Eliminating Terror: PM Tells Pak', *Daily Excelsior*, 28 October 2009.

As a substantial step in this direction, the border in Kashmir was partially opened. The official agreement was announced on 16 February 2005, when the Indian External Affairs Minister, K. Natwar Singh visited Pakistan and met the Pakistani President Pervez Musharraf at Rawalpindi. The contents of the agreement included: commencing the Thar Express train service, opening respective consulates in the cities of Mumbai and Karachi, and most importantly, starting an intra-Kashmir bus service between Srinagar and Muzaffarabad. India and Pakistan had differences over details of this new initiative. While India emphasized the use of passports as travel documents, Pakistan insisted on the disputed nature of Kashmir and advocated other mechanisms such as documents to be sanctioned by the United Nations or Red Cross, or by the local officials in the style of the old *Rahdari* system prevalent prior to 1953.[4] Finally, both the countries agreed on a formula by which the travellers, initially confined to members of divided families, had to carry a permit vetted by authorities from both the countries.

The Srinagar-Muzaffarabad road was the first intra-Kashmir route to open on 7 April 2005. Thirty members of divided families from either side of the LoC were allowed to take the fortnightly bus service termed the 'mother of all confidence building measures', a 'humanitarian measure', and 'a step towards bridging the divide in Kashmir'.[5] The commencement of the bus service between the winter capital of J&K, Srinagar, and Muzaffarabad, the capital of AJK, was applauded. The cumbersome process to facilitate travel started in March 2005. With regard to procedural norms, Indian citizens have to apply at the Regional Passport Office—the designated authority to evaluate applications, verify identities and issue entry permits after getting clearance from Pakistan. On 14 March, the distribution of travel forms started. Over 5,000 people gathered for 3,000 sanctioned travel forms to board the bus. Some 40 applicants were selected after the verification procedure,

[4] Under the *Rahdari* (transit) system, a letter from the local district commissioner was recognized as a travel document.

[5] For details on the Srinagar-Muzaffarabad bus see Debidatta Aurobinda Mahapatra, 'Bus Running for Reconciliation?', *Mainstream,* vol. 63, no. 20, 2005, pp. 27-9.

of which Pakistan cleared 29 names. These passengers, who were eagerly waiting to board the bus, had a close shave on 6 April, less than 18 hours before the scheduled departure, when a *fidayeen* (suicide bomber) attacked the Tourist Reception Centre at Srinagar where the passengers were staying. Eight of them cancelled their travel out of fear. Twenty-one passengers finally boarded the 35-seater bus flagged off by the Indian Prime Minister Manmohan Singh and the United Progressive Alliance (UPA) Chairperson Sonia Gandhi at the Sher-e-Kashmir Stadium. The vehicle, christened Caravan of Peace, carrying 'soldiers of peace' set off on a 170-km journey to reunite some of the divided families after decades of forced separation. Manmohan Singh, on this occasion, said, 'the Caravan of Peace is on its way, nothing can stop it now', thus renewing the hopes of thousands of divided families across the LoC. A tragic development took place on 8 October 2005, when an earthquake devastated many parts of Kashmir. The development also brought into focus the necessity of making borders soft. The earthquake prompted India and Pakistan to open five new foot crossings on the LoC on 29 October 2005, besides the Srinagar-Muzaffarabad route, to expedite relief operations and reunite divided families. These included Nowshera-Titwal, Chakoti-Uri, Hajipur-Uri, Rawalakote-Poonch, and Tattapani-Mendhar. Some of these meeting points witnessed a surge of thousands of people on both sides eager to meet their loved ones and find out their well-being. These meeting points were closed after a while. However, in order to continue the Caravan of Peace, another intra-Kashmir route between Poonch in J&K and Rawalakote in AJK was opened on 20 June 2006.

These developments beckoned a new era in the region as the earlier State-centric approach to conflict resolution was transcended by the people-centric approach. The porous border helped in myriad ways by addressing the grievances of the divided people to help bridge the artificial divide between them. The opening of the routes assisted in according fluidity to the border and provided an opportunity to revive the sentiments of belonging and fraternity shared by the people before the division. The opening of LoC was highly acclaimed by the international community. Then US

Ambassador to India, David Mulford, in a cable leaked by the Wikileaks observed,

> This is no ordinary road, and no ordinary bus; it is one of the biggest Kashmir stories in many, many years. . . . Although it does not deal directly with the 'core issue' of Kashmir, the bus has the potential to begin a new era in intra-Kashmir relations. Those who continue to dismiss it as a diversion or otherwise flawed sound increasingly cranky and out of touch with the march of history.[6]

THE REUNION

Opening of the routes have facilitated the reunification of many divided families. Hundreds of people have travelled through the two intra-Kashmir routes. According to Government of India figures, by the end of 2006, 3,000 people had travelled on the Srinagar-Muzaffarabad route, and 1,400 had used the Poonch-Rawalakote route. In January 2008, Taj Moiuddin, then Minister for Consumer Affairs and Public Distribution in the J&K government, stated that 5,786 people from divided families travelled along the opened Srinagar-Muzaffarabad road from April 2005 to 12 December 2007. Of these, 3,165 passengers from AJK travelled to J&K while 2,621 passengers crossed over to AJK from J&K.[7] According to official AJK sources, by the end of 2009, the bus lines had carried over 12,000 passengers.[8] Till December 2010, the number of passengers who had travelled to AJK on the Srinagar-Muzaffarabad route from the Indian side was 2,811 and from AJK, 4,043 to J&K. On the Poonch-Rawalakote route, 3,116 travelled from J&K and from AJK 4,833 passengers travelled to J&K.[9]

[6] 'Kashmiris First, Kashmir Later', *Kashmir Life*, vol. 3, no. 27, 12 September 2011. http://kashmirlife.net/index.php?option=com_content&view=article&id=1932:kashmiris-first-kashmir-later&catid=64:special-issue

[7] '786 members crossed over via Uri-Muzaffarabad border road'. http://www.indopia.in/India-usa-uk-news/latest-news/67107/National/1/20/1

[8] Shaheen Akhtar, 'Expanding Cross-LoC Interactions: Perspectives from Pakistan', Institute of Peace and Conflict Studies Issue Brief 130, September 2009. http://www.ipcs.org/pdf_file/issue/IB130-Ploughshares-Shaheen.pdf

[9] Ministry of Home Affairs, Government of India, *Annual Report, 2010-2011*, p. 11.

The narratives of people who have been able to reunite with their separated families are heart warning. Nasiruddin was one of those who boarded the first bus. He was less than two years old when he came to J&K with his mother to meet his aunt residing in Bijhama, a village in the district of Baramulla in the Kashmir Valley. They could not return due to the abrupt closure of roads following the division of Kashmir. 'My mother's last wish was to be buried in Muzaffarabad. It could not happen. When I went on Hajj, my one and only prayer was that I should be able to travel back to Muzaffarabad to meet relatives, most of whom I have never met', he said. Many of his fellow villagers, who could not get the permit to board the bus, gave him things to carry—a letter, a parcel or merely a message of a few words for their separated relatives.[10] All the passengers who have travelled via the two routes from either side of Kashmir have similar stories to recount. Setu Ram, a resident of Surankote (about 30 km from Poonch towards Jammu), originally belonging to Kotli in AJK, narrated how his heart-rending experience of partition was completely overshadowed during his trip to the other side where he could meet his deceased brother's family. In his words,

> I always cursed the governments of India and Pakistan for dividing our families and forgetting. . . . My brother is not alive. . . . When I met my nephew and his family, it was like a dream. They have converted to Islam though we continue to be Hindus. But religion did not come in. I forgave all those involved in dividing us. But I wish we could reunite permanently.

NOT A SMOOTH JOURNEY

The past experiences make it difficult for people to believe in the durability of the peace process, which in turn makes them desperate to visit their relatives. Interaction with people has brought to light scores of hurdles that have kept their dreams of reunification unfulfilled. One of the ironies related to the bus service is that

[10] Seema Shekhawat and Debidatta Aurobinda Mahapatra, *Contested Border and Division of Families in Kashmir: Contextualizing the Ordeal of the Kargil Women*, New Delhi: WISCOMP, 2009, pp. 31-2.

while the authorities receive thousands of applications, the buses run below their full capacity. The journey is not smooth in terms of bureaucratic hurdles ranging from complicated paper work to a wait of months for clearance. The procedure is lengthy and complicated, involving not only filling of forms, but also verification by the Criminal Investigation Department, counter-verification and clearance from both the countries. The patience of many runs out as the complicated permit procedure takes months and at times beyond that if a single formality is left out. Applications are often rejected without any clear explanation or are left pending for months. The AJK government between 7 April 2005 and 30 August 2007 dispatched a total of 4,114 applications to the J&K government for clearance to travel to Srinagar through the Chokothi-Uri crossing point. But only 1,254 applications were approved. Similarly, of the 3,938 applications received by the AJK government from J&K, only 1,070 were cleared.

There are many people whose dreams of reunion remained unfulfilled. Seventy-eight year old Nafisa Bi died in March 2007 waiting for the bus permit after having applied eight months earlier. Also, only those with close relatives living on the other side are eligible to apply and that has prevented many people to cross the divide and visit their native place where either their parents spent their lives or they have spent their childhood and youth. It needs to be emphasized that after the 1947 Indo-Pak War, most of the Hindu and Sikh communities had migrated to the Indian side of J&K, leaving behind their ancestral homes. Settled across the Jammu region, these communities are eager to visit their ancestral village and relatives living on other side of the LoC. Since some of these families do not have relatives across LoC but ancestral homes, it becomes difficult for them to get the permit to travel. Provisions need to be made for these people. Nostalgia runs across hundreds of narratives of these people who have no relatives on the other side but want to visit the place which was once their home. Haji Khuddam was 14 when he migrated to Bhimber in AJK from J&K with his family in 1947. The years have not dimmed Khuddam's childhood memories. He is still keen to visit his birthplace, but since he does not have relatives in the area, he cannot undertake the much cherished journey.

During the Second Round Table Conference between New Delhi and Srinagar, convened to generate a debate and evolve a consensus on various issues related to J&K, in May 2006, the Indian prime minister announced the establishment of five working groups responsible for looking into various contentious issues. These included confidence building measures (CBMs) across segments of society in the state, strengthening the intra-Kashmir links, economic development of the state, good governance, and strengthening relations between New Delhi and J&K. From the point of view of the divided families, the recommendations of the second working group are crucial. The group observed that the current travel procedures across the LoC are complicated and dilatory. It was also critical of the detailed scrutiny of applications that stretched over several months before the travel permit is granted. The group recommended necessary guidelines to simplify and speed up the procedure.

It stipulated that all residents of J&K would be required to obtain a Permanent Resident Certificate which is issued after extensive verifications. This may be taken as the basic document on the strength of which travel permit is issued. On production of the Permanent Resident Certificate, security clearance should be completed without fail within a period of two months by the authorities and the result communicated to the applicant. In the case of traders, the permit should be of multiple entry/ exit and valid for a minimum period of one year. In case of an emergency, like serious illness, death or marriage in the family, the permission should be granted on priority basis. Applications for travel permit may be accepted and disposed of at designated offices at Anantnag, Baramulla, Poonch and Rajouri in addition to the existing designated offices at Jammu and Srinagar. In the light of experience, the list of designated offices can be expanded further. The relatives of the travellers should be allowed to go up to the border check-post to see them off and also to receive the returning persons or those arriving from AJK at the border post. The recommendations also included expanding the category of persons eligible to travel through these cross-LoC routes such as persons wanting to visit places of religious interest or those who want to visit various places for tourism and persons in need of medical aid, provided they carry the requisite documents to show that they genuinely require treatment. Exchanges of groups of journalists, academicians, lawyers and any other groups which may be identified after due consultation may be authorized by the government. The universities and colleges in J&K may be authorized to organize exchanges of visits between students and faculty members on both sides of the LoC. Visits of school students from both government and private institutions may also be organized. The grant of admission in certain specific courses on a regular basis to AJK students in J&K universities may be considered. Cultural trips in the field of music, dance and drama and shows organized at designated places may be encouraged.[11]

In its meeting on 18 July 2008, the bilateral Working Group on Cross-LoC Measures set up by India and Pakistan agreed upon various steps to make the journey smooth, which included increase in the frequency of the Srinagar-Muzaffarabad and Poonch-Rawalakote bus services from fortnightly to weekly, introduction of a triple entry travel permit, expediting the processing of travel requests by allowing local authorities in Poonch and Rawalakote to receive applications previously handled only in Muzaffarabad and Srinagar and exchange of application forms through e-mail to expedite clearance.[12] This was followed by a meeting between the Foreign Secretaries of India and Pakistan on 26 July 2011. Later, the Minister of External Affairs of India, S.M. Krishna and the Minister of Foreign Affairs of Pakistan, Hina Rabbani Khar, met in New Delhi on 27 July 2011. In the meeting, the following crucial decisions were taken to facilitate cross-LoC travel:

- Cross-LoC travel would be expanded on both sides to include visits for tourism and religious pilgrimage. In this regard, the modalities will be worked out by both sides.
- Facilities including waiting area, terminal and clearing pro-

[11] 'Strengthening Relations across the Line of Control', Report of the Working Group II, January 2007. http://www.hinduonnet.com/nic/jk/jkreport_2.pdf.

[12] 'India, Pakistan Discuss Measures to Boost Trade, Transport'. http://www.thaindian.com/newsportal/south-asia/india-pakistan-discuss-measures-to-boost-trade-transport_10073228.html

cedures at the operational crossing points will be streamlined by both sides for smooth cross-LoC travel.

- The cross-LoC bus service between Srinagar-Muzaffarabad and Poonch-Rawalakote routes will henceforth run on every Monday.
- Application forms and requisite documentation in respect of travel across LoC will be exchanged by email between designated authorities of both sides. Such email transfer of application forms will be backed up by hard copies.
- Both sides will expedite the processing time for applications, which shall not be more than 45 days.
- Six-month multiple entry cross-LoC travel permits will be allowed by the designated authorities after completion of the required formalities at an early date.
- Coordination meetings between the designated authorities will be held at the terminals alternately on both sides of the LoC every quarter or as and when deemed necessary.[13] The fortnightly bus service on the Srinagar-Muzaffarabad and Poonch-Rawalakote routes were earlier converted into weekly bus service with effect from 11 September 2008 and 8 September 2008 respectively.

The chance to meet relatives after decades of separation has raised the expectations of the divided families. They now want permission for longer period of stay and multiple entry permits in order to meet emergencies and be liberated from the ordeal of repeating the complicated procedure. There is further need to take steps like making the travel across LoC hassle-free. Practices like change of bus at the LoC crossing point need to be given up. At a later stage, the permit system should be abolished and unrestricted movement allowed. There are people who want more than a temporary reunion or a single visit to their native places. Towards these goals, making the border immaterial is crucial.

[13] Joint Statement by S.M. Krishna and Hina Rabbani Khar, *The Hindu*, 27 July 2011.

CHAPTER 4

Soft Border: Economic Implications

The economic dimension of opening of cross-LoC routes is quite significant as it has led to prospects of intra-Kashmir trade and economic development. This move can prove to be advantageous since both sides possess enormous potential for cooperation, and in many ways complement each other. Two statements are of particular relevance in this context. On 31 December 2006, then Chief Minister of J&K, Ghulam Nabi Azad stated that some sort of 'joint management' with Kashmir across the LoC is possible in the fields of tourism, trade and water resources, which could pave the way for a lasting solution to the Kashmir issue. On 10 February 2007, Sardar Attique Khan, then prime minister of AJK offered to supply gas and electricity to the Indian part of Kashmir and invited entrepreneurs from 'all regions and religions' to invest in AJK. He also invited doctors and engineers from J&K and suggested that universities in the region admit students from AJK. The intra-Kashmir roads are well connected and can facilitate travel and trade between the two regions as well as between India and Pakistan. This chapter focuses on economic advantages of opening of cross-LoC routes. It also analyses the constraints faced by intra-Kashmir traders.

SETTING THE STAGE

The working group set up by the government of India on promoting intra-Kashmir linkages besides recommending people-to-people contact, as discussed in Chapter 3, also argued for initiating trade and commerce in its report submitted in January 2007. It said, 'The opening up of routes for trade and commerce would help in not only improving the economic conditions on both the sides but would also be an important step towards normalization of the situation.'[1] The group recommended a series of measures to facilitate trade. It argued that the opening up of trade and commerce between the two sides would lead to increased economic activity, new employment opportunities and help wean youth away from terrorism and other negative influences. The suggested measures included:[2]

- A joint consultative machinery of officials and representatives of trade and commerce (including the Chambers of Commerce) from both sides be formed to resolve any difficulties in the flow of trade and commerce.
- The list of items for export/import should include, *inter-alia*, handicrafts, fruits, cement, iron, medicines, sugar, basmati rice, machinery and equipment, sports goods including raw material like willow clefts, plastic pipes and any other items that are manufactured or assembled by small-scale industries in J&K.
- For import, the concerned chambers/traders of AJK should be asked to provide the list that can be finalized in consultation with their counterparts in J&K.
- The goods should be allowed to be transported in special vehicles authorized to travel across the LoC directly to the destinations for which the goods are intended. In case there are initial difficulties in this regard, transport depots at the LoC should be earmarked.
- It should be ensured that unloading, checking and loading is done quickly ensuring that the goods are not damaged in the process.
- Since exchange of goods is not expected to be very large, customs duty may be initially exempted on a reciprocal basis.
- Both sides should organize trade fairs at prominent places as also on a small-scale at some agreed border points.

[1] 'Strengthening Relations across the Line of Control', *Report of the Working Group II*, January 2007. http://www.hinduonnet.com/nic/jk/jkreport_2.pdf.

[2] Ibid.

- The roads on which goods and passenger traffic will take place may be widened and resurfaced to better specifications. The ultimate aim would be to encourage, in stages, the creation of a Free Trade Area comprising J&K and Pakistan-controlled Kashmir. Modalities for the same are being worked out between the two sides.

The recommendations paved the way for Indo-Pak trade diplomacy in 2008. In March 2008, Jairam Ramesh, the then Indian Minister for Commerce, announced that cross-LoC trade would commence in the near future. In May 2008, the issue was discussed during the visit of then Indian Foreign Minister, Pranab Mukherjee to Islamabad. In September 2008, on the sidelines of the UN General Assembly session in New York, Indian Prime Minister Manmohan Singh and Pakistani President Asif Ali Zardari agreed to allow cross-LoC trade in commonly agreed products. The Indo-Pak Joint Working Group on cross-LoC confidence building measures finalized the terms and conditions for trade on 22 September 2008. It cleared a list of 21 categories of items for duty-free trade. The list included: carpets, rugs, wall hangings, shawls and stoles, *Namdas* (a variety of Kashmiri carpet), *Gabbas* (a variety of Kashmiri carpet), embroidery items, furniture including walnut furniture, wooden handicrafts, saffron, aromatic plants, coriander, lentils, tamarind, black mushrooms, Kashmiri spices, kidney beans, honey, rice, precious stones, Peshawari leather sandals, medicinal herbs, maize and maize products, foam mattresses, cushions and pillows.

TRADE

On 21 October 2008 customs duty exempted trade began with the trading centres operating twice a week. On that day, 28 trucks loaded with goods crossed the Aman Setu (Peace Bridge)—13 drove across from the Indian side carrying apples and spices and 15 pick-up vans loaded with rice, spices, rock salt, dry dates and raisins drove to J&K from AJK. On the Poonch-Rawalakote route, three trucks carrying goods crossed over to AJK from the

Indian side, while three vehicles carrying goods from AJK entered J&K.[3] The J&K Governor N.N. Vohra flagged off the caravan of 13 trucks on the Srinagar-Muzaffarabad road. Describing the cross-LoC trade as an important confidence-building measure aimed at normalizing India-Pakistan relations, Vohra said the people of the region would benefit as the trade links would open up new vistas of economic progress.[4] At the Chakan-da-Bagh crossing point on Poonch-Rawalakote road, the Governor's Advisor H.H. Tyabji flagged off the first consignment of trucks. The Government of AJK declared the first day of trade as the 'Day of Commerce Caravan'. After that AJK Prime Minister Sardar Attique Ahmed Khan inaugurated the trading centre at Chakothi. He expressed hope that cross-LoC trade would help aid the resolution of the Kashmir problem.[5]

Since October 2008, when cross-LoC trade begun, the total volume of trade through the two points amounted to INR 2721.3 million in March 2010. The imports stood at INR 1602.1 million, while exports constituted INR 1119.2 million. Goods worth INR 1173.9 million were traded via Srinagar-Muzaffarabad route. The contribution of Poonch-Rawalakote was valued at INR 1547.4 million.[6] Till 31 December 2010, 4,434 trucks have crossed over to AJK from J&K and 4,477 trucks have crossed over to J&K from AJK through the two cross-LoC routes.[7] In April 2011, Pakistani President Asif Ali Zardari, while describing the cross-LoC trade between the two sides of Kashmir as a 'pioneering' confidence-building measure, informed that the trade has been 'growing in volume and value' and so far goods worth about INR 8,000 million have been traded.[8] On 20 September 2011, goods worth

[3] Ministry of Home Affairs, Government of India, *Annual Report, 2010-2011*, p. 11.

[4] Shujaat Bukhari, 'Apples, Honey Cross LoC', *The Hindu*, 22 October 2008.

[5] Ibid.

[6] 'Jammu & Kashmir Requests Centre to Increase Cross-LoC Trade Period', *Daily News and Analysis*, 22 March 2010.

[7] Ministry of Home Affairs, Government of India, *Annual Report, 2010-2011*, p. 11.

[8] 'Cross LoC Trade Pioneering CBM: Zardari', *Greater Kashmir*, 10 April 2011.

INR 34.5 million were traded at Chakan-da-Bagh crossing point in Poonch district. On this day, as many as 25 trucks carrying items such as red chilli, tamarind worth INR 81,82,345 rolled out from Trade Facilitation Centre in Poonch district to AJK. From AJK, 23 trucks loaded with almond, dry grapes, dates, herbs and Peshawari sandals worth INR 2,63,54,961 crossed over to J&K.[9]

The onset of travel and trade relations between the two parts of Kashmir is the beginning of a process of economic collaboration towards making the LoC irrelevant without any restrictions on the movement of men and material. If facilitated wholeheartedly, the trade can be significantly profitable not only for the two parts of Kashmir, but also for India and Pakistan. If the fundamental issues related to infrastructural bottlenecks can be overcome, the results can be highly beneficial. A restriction-free economic collaboration can generate the much needed interdependence necessary to bring in stability in the region and usher in a peaceful era in Kashmir. Increased intra-Kashmir economic activities can play a crucial role in transcending the politics of conflict with people rather than governments at the centre of transition.

CONSTRAINTS

When trade commenced between the two parts of Kashmir, many stakeholders in the peace process hoped that the 'humble beginning'[10] would help 'bridge the divide between the two sides'.[11] Mubeen Shah, President of the Kashmir Chamber of

[9] 'Over INR 34.5 Million Worth Goods Traded by J&K, POK Businessmen', IBNLive, 20 September 2011. http://ibnlive.in.com/generalnewsfeed/news/over-rs-345-cr-goods-traded-by-jk-pok-businessmen/830324.html

[10] At the day of commencement of trade, Governor N.N. Vohra stated it was an important day for people across LoC as the opening of the historic Jhelum-Valley Road for trade would cement relations between people across LoC. He said although this is a humble beginning, it is going to be a step towards the revival of the economy. To quote him, 'Though it would be symbolic in the beginning but we hope that the activity would also grow on the routes in future.' *Greater Kashmir*, 22 October 2008.

[11] Ibid.

Commerce and Industry who led a delegation to the Peace Bridge on 21 October 2008 observed, 'Today, it is a symbolic beginning. But I am sure we will manage to break all impediments and make this trade a reality for future generations.'[12] But the trade potential has not been fully realized as it is subject to various controls and hurdles. The trade was confined to only 21 items locally produced and manufactured from each side. The process of trade is complicated at times by unilateral alteration of the list. In 2009, India deleted garlic and dates from the list of tradable items, while Pakistan deleted lentils. Another problem is related to the placement of items in the list, some of which are available at a cheaper rate in the local markets. Restrictions such as limitations on days and timing, and issues of weight and transportation have constrained the trade links. Initially, trade was conducted on two days of the week—Tuesdays and Wednesdays from 9 a.m. to 4 p.m. A truck was allowed to carry only 1.5 metric tonne load. The goods were to be unloaded at the crossing point to be picked up by another local vehicle from across the LoC. Lack of proper storage facilities at the crossing points force traders to keep their goods under open sky until arrangements for local trucks are made. These problems raised 'the fuel, handling, freight, insurance, and other miscellaneous costs', seriously denting prospects of profit from trade at the LoC.[13]

The cross-LoC trade was supposed to be a step towards transforming the conflict in Kashmir but the deal is fast turning sour owing to the lack of a formal financial system for trade. The current barter system of trade is regulated by a joint chamber of commerce consisting of trade representatives from both the regions. The currency issue has not been resolved by the two countries even after three years of commencement of trade. This has become a major hurdle in expanding the LoC trade. The barter system with goods exchanged in comparable value is shrouded with vagueness and plagued by disagreements. In 2008, when trade commenced,

[12] *Indian Express*, 22 October 2008.

[13] Moeed Yusuf, 'Promoting Cross-LoC Trade in Kashmir: An Analysis of the Joint Chamber', U.S. Institute of Peace, special report, August 2009, p. 8.

traders from J&K who were engaged in fruit trade suspended transportation of fruits across the LoC owing to their perception that the exported products are underrated.[14] To quote a trader,

> The medieval barter system is playing havoc with our business as there is no acceptable method to get equal value for goods.... Courtesy lack of proper communication facilities and visa restrictions, we trade blindly with no knowledge about the market realities across the LoC.[15]

Asif Lone, a cross-LoC trader from Baramulla in North Kashmir observed, 'Every week you will see a newcomer in the trade. After facing serious problems, they retreat. . . . There is no guarantee of recovery of money if you do not know your trading partner across the LoC.'[16] Abdul Hamid of the Poonch Trade Facilitation Centre says, 'The efforts to resolve the (financial) dispute ended in fights. Traders exchanged blows and invectives. It was embarrassing for all.'[17]

As trade has increased manifold, it has become difficult to carry on with the barter system. The demand for facilitating financial transactions for trading across the LoC is growing. Kashmir Valley's apex industrial body, the Federation of Chamber of Industries in Kashmir, has proposed that trade should be conducted in dual currency through 'opening of Nostro and Vostro accounts by J&K Bank and some designated bank from AJK in Srinagar and Muzaffarabad'.[18] In September 2011, the former prime minister of AJK, Sultan Mehmood Chaudhary, on a five-day unofficial visit to J&K, advocated formation of a new joint chamber of traders. He argued,

[14] 'Fruit Growers in J-K Suspend Cross Border Trade with Pakistan', *The Indian*, 13 November 2008. www.thaindian.com/newsportal/india-news/fruit-growers-in-j-k-suspend-cross-border-trade-with-pakistan_100118757.html.

[15] Personal Interview, 20 February 2011.

[16] 'Mistrust Prevails in Cross-LOC Trading', *Business Recorder*, 17 August 2011, http://www.brecorder.com/pakistan/business-aeconomy/24091-mistrust-prevails-in-cross-loc-trading.html

[17] Ibid.

[18] 'FCIK Demands Inclusion of Services in Cross-LoC trade', *Greater Kashmir*, 22 September 2011.

I think this is where the solution to your (business community) problems lies. When the traders of the two sides meet, it will go a long way to help resolve the issues of traders of the two sides. But one major thing is needed for that—the LoC should be made irrelevant and there should be free movement across the LoC.[19]

But the suggestions or demands of the business community get only occasional hearing and most are unable to put pressure on the respective governments, individually or even jointly. Local people or local authorities have no decision-making power to change the contours of cross-LoC trade, which subjects this novel mechanism for peace to the approaches of power centres at New Delhi and Islamabad. The existing mechanism of the Joint Chamber of Commerce and Industry to facilitate cross-LoC trade is a toothless tiger without significant power to decide on matters of trade.

Stringent visa rules restrict exploration of markets and deprive traders the opportunity to meet their counterparts regularly and personally across the LoC, thus casting a negative spell on trade prospects. Communication even through telephone is not easy. It needs mention that this facility was completely banned ostensibly due to militancy in J&K. In 2010, in a laudable move to make it possible for the traders from J&K to communicate with their counterparts in AJK, the Telecommunications Ministry of India provided ISD facilities to Pakistan on a few specific telephone numbers in J&K. Nearly after two decades, the Indian government, in April 2010, started a direct phone link with AJK to facilitate trade. Nevertheless, the J&K traders suspended trade during April and May 2011 to pressurize the government to look into their demands that go beyond limited communication. There are a couple of issues the redressal of which can facilitate expansion of trade, the foremost being the financial aspect. On 26 August 2011, while the J&K legislature was not in session, Governor N.N. Vohra, on the recommendations of Chief Minister Omar Abdullah, exercised powers conferred on him under Section 91 of the Constitution of J&K by promulgating an ordinance with

[19] 'Make LoC Irrelevant, says former PoK Premier', *The Hindu*, 22 September 2011.

implications for cross-LoC trade. The ordinance, which came into immediate effect, introduced amendments to the Value Added Tax (VAT) Act, 2005 to boost cross-LoC trade by marking application of zero VAT on goods locally produced and manufactured and sold to the traders across the LoC via the two routes. But traders opine that the move meant to foster better trade has, however, created confusion in the absence of banking facilities, bureaucratic hurdles and infrastructure problems.

Mutual suspicion and inflated security concerns have impacted prospects of trade. On 10 February 2011, the authorities from AJK requested their J&K counterparts to limit the number of trucks crossing over to AJK to 50 per day on the Srinagar-Muzaffarabad route. The limit was later extended to 100 trucks. The average number of trucks crossing the peace bridge on the LoC is 200 per week but occasionally, the number crosses even 300. The Indian army demanded that the number of trucks be limited a month after a similar request from AJK. The reason put forward was security concerns since security personnel have to check all the trucks that cross the LoC thoroughly and this is a time-consuming process. The cross-LoC trade between the divided parts of Kashmir has been growing in volume and value. Such limitations imposed on cross-LoC trade may likely discourage traders to increase trade, ultimately affecting the prospects of economic development in the region.

During the meeting of foreign ministers of India and Pakistan in New Delhi on 27 July 2011, many crucial decisions were taken to increase trade across the LoC. These include:

- List of 21 products of permissible items for cross-LoC trade will be respected by both sides. The working group will review the trading list with a view to further specifying permissible items to facilitate intra-Kashmir trade.
- Both sides will provide adequate facilities at the trade facilitation centres on each side.
- The number of trading days was extended from two to four per week. Truck movements shall take place on Tuesdays, Wednesdays, Thursdays and Fridays, both on Srinagar-Muzaffarabad and Poonch-Rawalakote routes.

- The designated authorities will resolve operational issues concerning cross-LoC trade through regular interaction.
- Regular meetings between the Chambers of Commerce and traders of both sides will be facilitated.
- Existing telephone communication facilities should be strengthened.
- The meetings of the designated authorities will be held alternately at the Crossing Point terminals on both sides of the LoC every quarter or as and when deemed necessary.[20]

The Joint Working Group on Cross-LoC CBMs established in 2008 remains largely inoperative as it does not have any power to implement its recommendations. As a step to promote intra-Kashmir trade, India and Pakistan agreed in July 2011 that the Joint Working Group would meet on a bi-annual basis to review existing arrangements and suggest additional measures for cross-LoC travel and trade. This is a welcome step as it keeps hopes alive that the demands of the traders and travellers will be met and the border will further soften.

Many demands towards promoting cross-LoC trade such as increasing the number of items and multiple entry visas are still pending. Steps are required to widen the scope of trade so that all the commodities produced or manufactured on either side of the LoC are allowed for trade. Besides trade in goods and commodities, facilities should also be extended to services, joint ventures to explore local resources. From a larger perspective, these joint ventures can moderate much of the controversial nature of projects undertaken on either side. The hydel power projects in Kashmir often become controversial between India and Pakistan and, therefore, joint venture projects can be explored. There are huge mining reserves in the LoC area and joint venture projects can lead to mutual profits.[21] The lack of power of the people in whose name measures were initiated to make it a soft border induces a sense of pessimism in the minds of people including traders, who

[20] Joint Statement by S.M. Krishna and Hina Rabbani Khar, *The Hindu*, 27 July 2011.

[21] 'FCIK Demands Inclusion of Services in Cross-LoC Trade', *Greater Kashmir*, 22 September 2011.

want more involvement in the decision making process to facilitate cross-LoC linkages. This sense of involvement will definitely add to the dynamism of the peace process.

CHAPTER 5

Opening Traditional Routes

As the peace process gathered momentum, the focus seemed to turn towards people, rather than political claims of either country, implying that it is people who must be given the opportunity beyond political bracketing to shape their destiny. In a historic gesture, two intra-Kashmir routes were opened in 2005 and 2006. Taking further steps in terms of facilitating trade and travel through these routes, as discussed in earlier chapters, can be quite fruitful. The process of making Kashmir borderless cannot stop at the peace measures taken so far. The prospects of opening other intra-Kashmir routes appear feasible as opening of the two routes brought into focus the humanitarian and economic advantages of opening all the routes and also their political implications in terms of softening of positions of India and Pakistan. Opening traditional routes, which played a crucial role in the lives of the people of the region before the division, has currently emerged vital in the peace discourse. The need to move beyond symbolic gestures and look into the prospects of bringing further flexibility into the border by opening all intra-Kashmir routes has been increasingly felt by the people, who in turn pressurized their political leaders to take the issue in right earnest. This chapter argues that there is a need to move beyond the opening of two intra-Kashmir routes and instead open other traditional routes since it will enable the peace process to focus on people-centric resolution of the conflict. It focuses on some of the traditional routes and their prospects towards enriching lives and bringing back to them the old shared identity. It will emphasize how opening of these routes will bring economic advantages to the people as well as the states, and eventually propel

India and Pakistan to think out of the box, keeping aside their rigid positions towards transformation of conflict in Kashmir.

THE ROUTES

The years from 2005 to 2007 witnessed the visit of five civil society delegations from the Pakistani side of Kashmir to J&K, by the intra-Kashmir bus services. The mutual interactions of these people from both sides of the divide provided an opportunity for them to think independently, separately from the standpoint of their respective governments. The talks in Jammu (September 2005) and New Delhi (September 2005 and May 2007) provided the occasion for leaders and activists from J&K, AJK and Gilgit-Baltistan to express themselves on a single platform, to know each other and revive old sentiments of belongingness and fraternity. One common theme that emerged during interactions was the necessity to open traditional routes in a region which had witnessed unrestricted flow of men and material. Before 1947, dozens of land routes were operational in Kashmir and used by people to travel for trade as well as to meet their relatives scattered all over the region. They include the following identified routes, as recommended by the 2nd Working Group set-up by the Indian prime minister to promote cross-border relations: Kargil-Skardu, Turtuk-Khapulu, Gurez-Astore-Gilgit, Jhangar (Nowshera)-Mirpur and Kotli, Jammu-Sialkote Chhamb-Jorian-Mirpur, and Titwal-Chilhan.

BRANCHES OF THE SILK ROUTE

Some of the traditional routes in undivided Kashmir were part of the grand Silk Route. The term 'silk route' is a generic term used in a broader sense to define the routes and their branches in Eurasian region that witnessed the traverse of ideas and goods in ancient as well as medieval periods and later till the rigid lines across states were demarcated. Prior to 1947, Kashmir was a major bridge between India, China and Central Asian countries as some of the branches of the Silk Route passed through it. Notwithstanding the rugged terrain and apparent remoteness, merchants, explorers, spies, soldiers, missionaries as well as adventurers traversed the region. The Silk Route via Kashmir played a seminal role in transporting not only commodities for trade, but also knowledge and ideas. The flow of Buddhism from India through Central Asia to China and the flow of Sufism from Central Asia to India are two major examples of this phenomenon.[1] Around 217 BC, Kashmir emerged as a transit route for the spread of Buddhism to China. Ashoka of the Maurya Empire and Kanishka of the Kushan Empire played crucial roles in spreading the Buddhist tenets in the region. It is believed that the 4th Buddhist Conference was held in Kashmir. Sufism, a moderate branch of Islam, with major emphasis on benevolence and tolerance came to Kashmir and other parts of India through this route. Though Islam entered the Indian subcontinent in the seventh century, its popularity can be attributed to the Sufi saints, many of whom came to India from Bukhara, Samarkand and other Central Asian cities.[2]

There are three major branches, which can be subdivided into sub-routes of the Silk Route that traverses Kashmir. Mushtaq A. Kaw has detailed these branches in the following way: One such route originating from the Punjab, reached Srinagar, then moved ahead to Ladakh in J&K, where it split into two branches, one, accessed the Tibetan part of China via Chishool, and another, via Nubra, crossed over the Karakoram at Daulatbeg Ulde and reached Xinjiang Uighur Autonomous Republic, from where it moved further towards Central Asia. This is known as the Ladakh route. The second major route from the Punjab passed over Srinagar, Gurais and Bandipora and entered Gilgit in the northern Areas of Pakistan where it fragmented into three sub-branches. The first one took an easterly direction and terminated at Xinjiang across Karakoram; the second and third took westerly directions, crossed over the Pamirs and reached Tajikistan across Murghab and Badakhshan

[1] For Kashmir's Central Asian connections through the Silk Route see Debidatta Aurobinda Mahapatra, *Central Eurasia: Geopolitics, Compulsions and Connections, Factoring India*, New Delhi: Lancers Books, 2008, pp. 149-55.

[2] Nirmala Joshi, 'Introduction', in Nirmala Joshi, ed., *Reconnecting India and Central Asia: Emerging Security and Economic Dimensions*, New Delhi: Pentagon Press, 2011, pp. 21-3.

on the one hand, and Wakhan Corridor in the Afghanistan-controlled Badakhshan region on the other. On reaching Balakh, the Wakhan route split into two branches—one moved towards Kabul and another towards Tajikistan, Uzbekistan, Turkmenistan and Iran at Kunduz, Termiz, Mazar-i-Sharif, Jalalabad, Sheberghan and Meshad respectively. The third principal route stretched from Srinagar to Central Asia and Iran via Baramulla, Uri, Muzaffarabad, Abbotabad, Rawalpindi, Peshawar, Kabul and Afghanistan across the Hindukush and Khyber passes.[3]

These major routes are interconnected by a network of smaller routes such as Kargil-Skardu, which was one of the major traditional routes in the Ladakh region. This author's detailed study of this route reveals that in order to access the route from the Indian mainland,[4] one has to travel from Srinagar towards Kargil town situated at a distance of 205 km. Before 6 km from Kargil town, one has to turn north by crossing the Singo river on a route closed for civilians except the local borderlanders with identity cards. Then, alongside the river, one has to travel about 6 km to the last post of the Indian army. Though the movement of people is restricted beyond this point, the route moves onwards towards Baltistan. From the last post of the Indian army, Skardu is about 169 km and can be reached within six hours. The opening of this route has remained in news since 2006 when the Poonch-Rawalakote road was opened. The people of Kargil hoped they would benefit from what the former president of Pakistan, General Pervez Musharraf, called a 'fleeting moment' of history which needed to be capitalized as those moments would not recur. While interacting with a group of people from the border areas of Kargil, then Governor of J&K, S.K. Sinha said that he had taken up the issue of this vital link road with the Indian prime minister. In June 2005, during his visit to Kargil to lay the foundation stone for a

[3] Mushtaq A. Kaw, 'Restoring India's Silk Route Links with South and Central Asia across Kashmir: Challenges and Opportunities', *The China and Eurasia Forum Quarterly*, vol. 7, no. 2, p. 60.

[4] Debidatta Aurobinda Mahapatra, 'Silk Route in Kashmir', *Central Eurasian Studies Review*, vol. 8, no. 1, Summer 2009, pp. 13-15.

hydroelectric project, the Indian Prime Minister Manmohan Singh said,

> I have been told that the people of Kargil are keen on having the traditional links restored with Gilgit and Baltistan. I am aware that this is a traditional route through which trade was conducted between Ladakh and Kashmir for ages. We are examining a proposal to start talks with Pakistan to open the Kargil-Skardu road.[5]

Despite official talks, the route still remains rigid giving rise to a sense of disappointment that their demand, which for them was genuine, was trivial and inconsequential for the States on either side.

Gurez-Astore-Gilgit road was also part of the Silk Route, evident from the fact that archaeological surveys in the Gurez Valley, particularly in Chilas, have discovered a number of inscriptions recorded in stone in Kharoshthi, Brahmi and Tibetan languages. The Gurez-Astore-Gilgit route, starting from Bandipore in the Kashmir Valley, passes through the 11,672 ft high Razdhan pass also known as Sanga Top to descend to the picturesque Gurez Valley to proceed towards Gilgit via Bunji and Astore. It was operational till 1947. The road was frequently used by the British army to supply goods to its garrison in Gilgit. It was called the Gilgit Transport Road. Aurel Stein, in his book, *Ancient Khotan* details the route. To quote him,

> The 'Gilgit Transport Road', which I was authorized to use for the first portion of my journey, from Kashmir to Gilgit, dates in its present form only from the years 1890-92, when the placing of an imperial garrison in Gilgit and in the adjoining valleys leading towards the Eastern Hindukush necessitated the construction of a military road fit for laden transport during a certain time of the year. But the route which it follows, between and over the high ranges separating the Kashmir Valley from that of the Indus, is marked out by nature as the most accessible line of communication from Kashmir to the Dard territories northward, and there is historical evidence to prove its use at an early period.[6]

[5] Shujaat Bukhari, 'Talks for Opening Kargil, Skardu Road: Manmohan', *The Hindu*, 12 June 2005.

[6] Mohammad Ashraf, Gilgit Transport Road, 28 September 2008. http://www.kashmirfirst.com/articles/other/080928_gilgit_road.htm

The Gurez Valley is part of ancient Dardistan, now divided into three regions—Neelam district (from Baghtor to Sharda Peeth) in AJK in Pakistan, Astore (from Kamri to Minimarg) in Gilgit-Baltistan in Pakistan, and Gurez Valley (from Baghtor to Abdullae Tulail) in India. The Dard tribe, claiming original Aryan ancestry, inhabited the region since centuries but the tribe was divided with the partition of Kashmir. Turtuk-Khapulu to Skardu, from earliest times has been another crucial trade link. From Leh after hiking the Khardung-La, the world's highest motorable road at 18,380 feet, one can descend into the Nubra Valley where this road lies. The route starts near the Tyakshi village in the Turtuk sub-sector and following the left side of the river Shayok, passes through Drospinthang to proceed to Skardu in Gilgit-Baltistan, covering a distance of about 200 km.

BENEFITS

The opening of these routes can lead to the revival of the traditional Silk Route, thus facilitating the cross-LoC traffic not only for local people or India and Pakistan but also for Afghanistan, Central Asia, and Tibet and other parts of China such as Xinjiang, thereby restoring free flow of trade and travel that existed since ancient times. The closure of the Silk Route in Kashmir has almost a smothering impact on its people and its trade prospects since the region was cut-off from the rest of the world. The closure has left a few long and circuitous options for transport and communication. Reviving these traditional routes would undoubtedly put Kashmir back on the centrestage of the global map as a centre of the Silk Route trade, and also as a centre for historical learning. Kashmir can once again become a hub of trade, commerce and other activities. The opening of this route will not only provide a direct link to Central Asia, but also open up unexplored and unexploited potential of tourism, especially adventure tourism. Prior to the division, this whole region having all these routes was a popular destination for tourists. The older generation remembers the region as a 'trekkers' paradise'. The tourists can enjoy the beauty of the region as a whole if the road links are opened, which in return will bring economic prosperity.

The benefits for the local people and the region go beyond the advantages mentioned above, as the Silk Route's importance in hard mountainous climate and topography cannot be undermined. The utility of some branches of the route lies in their durability in winter months. The regions such as Kargil, Turtuk and Gurez remain largely cut off from other parts of J&K and consequently mainland India for about seven months every year due to inclement weather resulting in heavy snowfall thus bringing a plethora of problems for the residents ranging from shortage of food supplies to health emergencies. The opening of the above routes in these areas can provide the people of the region an all-weather road that remains pliable throughout the year. The Kargil-Skardu route that lies in a cold but arid region remains in pliable condition during the harsh winter months. The local people in Kargil can use this road in winters to travel to Skardu and other areas of Gilgit-Baltistan for shopping and other necessary purposes. The local people argue that this would bring economic benefits since from Skardu, apricot, raisins and medicinal herbs can be sold in Kargil, while from Kargil, sugar, tea, garments and vegetables can be traded. The Gilgit-Baltistan traders, in a memorandum submitted to the Government of Pakistan in January 2009 demanded the reopening of the Skardu-Kargil and Astore-Srinagar roads and access of Gilgit-Baltistan to Tajikistan via the Wakhan corridor to harness the trade potential of the area. Turtuk-Khapulu road can make a triangle of the all-weather road as the route can be easily joined to Skardu that can be further extended to Kargil, which can be linked to Turtuk via Nubra. All these linkages can play an effective role towards economic development of the region.

The Srinagar-Muzaffarabad route was the first one to be opened for the diplomatic capital it held and it was hyped to demonstrate the friendly relations between the two countries. The 'mother of all confidence-building measures' brought smiles to the faces of the thousands of divided families across the LoC since they hoped this symbolic gesture would pave the way for the opening of other intra-Kashmir routes in their areas. On the Poonch-Rawalakote road, the rush of people is still evident but the initial interest shown in the Srinagar-Muzaffarabad route has somewhat gone down as the Kashmir Valley does not have a large number of divided families. The people of Poonch-Rajouri in Jammu and those of Kargil and Leh in Ladakh share ethnic, emotional, linguistic, cultural and familial ties with the people living across the LoC. Half the passengers who came from AJK on 7 April 2005 via the Srinagar-Muzaffarabad route had relatives in the Poonch district of Jammu. They had to travel 170 km from Muzaffarabad to Srinagar, another 300 km from Srinagar to Jammu and again, 250 km from Jammu to Poonch to meet their relatives. They could have reached their destination within hours if the Poonch-Rawalakote route were open by that time, as through this route they would have been required to travel less than 200 km. Later, the Poonch-Rawalakote road was opened but the routes that were discussed above too are quite significant in terms of humanitarian considerations. Before 1947, Baltistan and Ladakh were part of a single entity, sharing a common geography, language, culture and history. There is palpable excitement among the Balti-speaking families that the opening of the Kargil-Skardu route will enable them to revive their common culture and facilitate the reunion of hundreds of divided families. Longing for opening of the less discussed Gurez-Gilgit route is acute among the divided families living in the region. Despite the division of what once was known as Dardistan in India and Pakistan, the Dard tribe, sharing common language (Shina), dressing patterns and culture with Gilgit-Baltistan, is hopeful that a time will come when their fragmented abode will reunite. The Turtuk region had five villages named Turtuk, Tyakshi, Thang, Pachathang and Chalunka in Pakistan controlled Kashmir till the 1971 war. The abrupt division and redrawing of the border after the war had again led to division of families longing for reunion. The reopening of these routes can play an effective role towards not only bringing the members of divided families together, but also help transform conflict in one of the most 'dangerous places of the world'.

OTHER ROUTES

Besides the Silk Route, there are many other routes mainly in the Jammu region whose opening can play an important role in enhancing peace prospects in the region. One such road is the Jhangar-Mirpur Road connecting the Rajouri belt of Jammu region with the Mirpur sector of AJK. Mirpur is just a few kilometres away from Jhangar—the last point on the LoC on the Nowshera border. The route is part of the old Mughal Road with a traceable history of about 600 years. It was used by the Mughal emperor Akbar and his successors. The Mughals from their centre of power in Agra and Delhi used to travel to picturesque Kashmir via Lahore, Mirpur, Rajouri and Shopian. En route, they had to cross the Pir Panjal mountain range separating regions of Jammu & Kashmir. On the way to Kashmir, the Mughals built numerous forts and rest houses, which still stand erect even after centuries of wear and tear. Some of the notable landmarks which this author came across during surveys in the region are Changez Sarai (where the intestine of Emperor Jehangir, who fell ill and died during one of his visits to Kashmir, is preserved) near Rajouri and Noori ki Chham (the bathing place of Noor Jahan, the consort of Jehangir) near Bufliaz in Poonch district.[7] Poonch was divided into two regions due to the India-Pakistan war and, thereby, thousands of people remained divided until the Poonch-Rawalakote route was opened in 2006. The demand for opening this Mughal route of historical significance is currently growing. It can further facilitate the reunion of divided families and also bring in economic benefits.

The Chhamb-Jorian-Mirpur route will reconnect the cities of Jammu to Mirpur, located at a distance of about 100 km. Jammu-Mirpur route via Chhamb-Jorian was frequented by travellers and traders since ancient times. This road on plain surface is altogether free of seasonal effects or problems associated with high mountain terrain and landslides. It is motorable throughout the year. This

[7] Debidatta Aurobinda Mahapatra, 'Poonch for Peace: An Eye Witness Account', *Kashmir Times*, 6 May 2007.

road link will lead to faster flow of traffic as it would involve just two-and-a-half hours to cover the distance from Chhamb to Mirpur. The road will connect a number of cities in Jammu including Akhnoor and Udhampur with cities in AJK such as Kotli. There are divided families and traders who are keen to see this road opened. The route also has significance in terms of enabling people to visit their native places which they were forced to leave in the aftermath of the partition of Kashmir. Thousands of the displaced from Mirpur, settled in Jammu, have renewed their demand for opening this route.

The 42-km Jammu-Sialkot road via Suchetgarh was operational before 1947. The road passes through the international border between India and Pakistan in Kashmir. There was also rail connection between Jammu to Sialkot. Pakistan has rejected the idea of opening this route ostensibly on the premise that the road is on the India-Pakistan international border, and not on LoC. Pakistan leadership appears to be only in favour of opening the routes across the LoC. Opening of the Jammu-Sialkot Road is important for divided families living in the Jammu plains as most of their relatives live in the Sialkot area—an hour's drive. Travel by Srinagar-Muzaffarabad or even Poonch-Rawalakote roads are time-consuming and costly for these people. Sialkot and Jammu also have historical and cultural ties that become evident during the annual festival of Baba Chamliyal Mela where people from both sides are allowed to join the event for a few hours. The opening of this particular road will bring economic benefits by providing local people and traders an opportunity to access markets on the other side. Farmers from the Jammu region can sell their produce (particularly the world-famous Basmati rice) on the other side of the LoC.

Opening the Titwal-Chilhan road on the Neelam river, which divides Teetwal sector in Kupwara district in J&K from Chillan in AJK, can facilitate cross-LoC traffic since the Neelam River marks the LoC in the area. Most of the families residing in this area have relatives on the other side of the LoC. On 26 January 2004, people from the two sides were allowed to gather on their respective sides on the banks of the Neelam to communicate from a distance.

Thousands of people grabbed this opportunity to at least see each other. There were even exchange of gifts and letters by slanging ropes across the river. The temporary reunion rekindled hopes to meet each other. Paradoxically, the realization to open cross-LoC routes such as Titwal-Chilhan emanated from a tragedy. On 8 October 2005 an earthquake shook the undivided Kashmir without recognizing borders as erected by the states. In the aftermath of the earthquake, India and Pakistan agreed on opening of five points on the LoC. Titwal-Chilhan was the first point opened for divided families on 20 November 2005. A relief camp was also set-up in Titwal where people from AJK could come for relief and aid. But the reunion was short-lived and the point was closed after a while.

The closure of the above-mentioned and many other routes doomed the life of the local people and fractured their livelihood and socio-cultural life patterns. Despite long years of negligence and non-use, most of these routes still remain intact and with minor repair costs can again be reopened. The entire Turtuk-Kharpulu route is motorable and only needs some patchwork or widening at certain points. At the last Indian army post, one can see the Kargil-Skardu route is blocked by a stone wall. All that needs to be done is to demolish the stone wall. When that historic event will take place remains to be seen. Opening all cross-LoC routes will help undivided Kashmir towards harnessing its economic potential and regain centrestage in terms of trade and also in terms of reversing, or at least soothing, the tragic history of separation and suffering by uniting the besieged families.

Conclusion

The main thrust of this study has been to explore the prospects of a borderless Kashmir, in which the conflict terminology, at least from a state perspective, will be defunct and a new borderless and peaceful and economically vibrant Kashmir will emerge on the scene of South Asia without actually denting but transcending the inflexible position of the two main rivals India and Pakistan. Both the countries, since their emergence from the shadows of British colonialism, have made Kashmir a crucial issue in bilateral relations with consequent bearing of their bilateral conducts on the lives of the people of the region, whose destiny they seek to resolve by focusing on state-centric mechanisms such as wars. It is no wonder then that India and Pakistan have fought many wars, suffered tragic losses on account of men and material, thus stunting prospects of development of these two developing countries and the region of Kashmir because the cost of maintaining the conflict has increased disproportionately over the years. As I have argued, a transformed borderless Kashmir will not only minimize the cost of conflict, but also bring into board the prospects of evolution of Kashmir from a 'dangerous place on earth' to a peaceful place and revive its status as 'paradise on earth'.

The history of the Kashmir conflict and its course since its inception in the late 1940s is a complicated narrative, which this study has not attempted to explore as it falls beyond its scope, but it needs emphasis that Kashmir is one of most intractable conflicts in the world. Hence, the change in the minds of leaders of India and Pakistan and leaders of Kashmir, as visible in recent years, towards fostering peace in the region is no small wonder. The post-Cold War developments and related changes in national and international politics impacted the politics of the region to make the leaders realize that wars and consequent drawing and redrawing of borders bring death and destruction in their trail, hence, in its place peace must be cultivated through innovative measures.

The emerging discipline of border studies brought into focus that borders are not mere lines of separation but also lines of suffering and fractured identity. In the case of Kashmir, the border, drawn abruptly and indiscriminately just on the basis of control of territory by respective States at the time of ceasefire, passed through villages, fields and even houses. While states announced victory or defeat after the wars, the people living in the contested spaces for centuries with united culture, languages and blood relations, suddenly found themselves separated by force with their lives becoming fragmented almost in an irrevocable manner. Families were fragmented, culture besieged, peaceful lives turned into lives of permanent anxiety, sources of livelihood compromised or jeopardized—these are a few of the consequences which the people living around the artificial divide had to confront. The saga of suffering does not end there. While States from their capitals of New Delhi and Islamabad plan and execute plans on a realist pedestal, it is the borderlanders who are lost in the maze of State policies. The irony then emerges—the states which erect lines of division to protect or provide security to its people, actually finds itself as the leviathan which sucks out life from the people to whom it claims to provide security and prosperity. This leads to the moot question, which the study has attempted to explore—can conflict in Kashmir be transcended by transforming the LoC into a line of contact, commerce and cooperation? Can borders be erased, rather than drawn and redrawn so that people can lead a normal life with no displacement, no fear of firing or shelling or stepping on a mine, no restriction on mobility and no forceful separation? Is it possible to erase those lines at least in spirit if not in letter so that they remain as markers of sovereignty of India and Pakistan on maps but nonexistent in terms of flow of people, goods and ideas?

It is the people and their aspirations to move closer, and their determination to see out of the prisms contrived by the state—in sum, the bottom-up movement without any impositions from state—that played a significant role towards softening of the divide. When India and Pakistan started the peace process, it was the people who provided the needed momentum and sustained the spirit of the peace process. Phrases such as 'heart-to-heart talks', or 'round tables', are not mere phrases created out of lexicography for hype but saw action in those years. As the author participated in a number of events where people from both sides met, their indomitable spirit and craving to be reunited was apparent. Emotions abound as husbands meet separated wives, or fathers meet sons, or individuals visiting their place of birth, as in the case of Ghulam Mohammad from AJK who following his heart disobeyed the permit rules and breathed his last in J&K. These are not only heart-rending narratives to be remembered and narrated; but they also bring in their trail stories of unremitting agony these people suffer without any fault of theirs. A borderless Kashmir, with faint mechanisms of border to satisfy national-ego, can assuage much of the sufferings of these people.

The developments in terms of opening of the Srinagar-Muzaffarabad and Poonch-Rawalakote routes have accorded a semblance of irreversibility to the peace process despite setbacks. The Mumbai attack of 2008 killed more than 150 people, both Indian citizens and foreigners, raised the pitch of animosity between India and Pakistan and affected the peace process, but it did not stifle the process absolutely as the two countries came back to the negotiating table after considerable sabre-rattling. As I have argued, peace constituency has not been confined to states and their officials; rather the peace momentum has been substantially backed up at the grassroots level. In other words, the peace substructure at the grassroots has upheld the peace superstructure at the state level. Despite ups and downs in the history of the Kashmir conflict and in the history of India-Pakistan relations, the realization has not dawned as starkly as in recent years that the peace momentum must be sustained towards resolving the complex issue. Towards this goal, making borders irrelevant is one of the basic necessities as well as a gigantic leap towards factoring people and their demands in the peace discourse. There are, no doubt, detractors of peace process on both the sides, but their constituency has shrunk with the passing of time.

The re-opening of the cross-LoC routes at two points has opened new opportunities for the local people. The advantages are certainly numerous from both humanitarian and economic points of view. There has been a desire swelling on the ground that all other routes must be re-opened. The humanitarian advantages are manifold: first, it will bring together the families who have been separated for decades; second, the people will get the much cherished opportunity to feel united with their brethren on other side of the division, to visit the other side of the immaterial border at their will and share feelings and emotions, towards an integrated identity ruptured for decades; and third, it will, by pushing the people's spirit towards peace and reducing the constituency of radicals who believe in methods of violence, transform Kashmir from a zone of conflict to a zone of peace. The trade benefits of an opened LoC are numerous since both Kashmirs complement each other in many ways: first, what is in opulence in one side is in scarcity on other side; second, transport and communication in Kashmir will be smoothened if the cross-LoC roads are re-opened; and third, intra-Kashmir trade can bring in development and prosperity for the people. The traditional routes are further linked beyond Kashmir to Central Asia and China, thus prospects of international trade and tourism cannot be ruled out. Kashmir, if one looks at its geography, is centrally located on the Silk Route trade with a cobweb of transport and communication routes existent since ancient times. Most of the routes are intact and pliable even after their closure for decades. This author's study of Kargil-Skardu route reveals that the route is completely motorable even during the harsh winter months, but the local people are deprived of the opportunity to use this road and suffer hardship as they have to travel long and circuitous routes to markets in Leh or Srinagar. There is no point in creating obstacles in opening these routes. India and Pakistan must moderate their policies in a move to listen to the people of the region rather than to hard core realism. A borderless Kashmir can usher in permanent and positive peace in one of the most violent regions of the world.

Bibliography

Akhtar, Shaheen, 'Expanding Cross-LoC Interactions: Perspectives from Pakistan', Institute of Peace and Conflict Studies Issue Brief 130, September 2009. http://www.ipcs.org/pdf_file/issue/IB130-Ploughshares-Shaheen.pdf

Anderson, Malcolm, *Frontiers: Territory and State Formation in the Modern World*, Malden: Polity Press, 1996.

Ashraf, Mohammad, 'Gilgit Transport Road', 28 September 2008. http://www.kashmirfirst.com/articles/other/080928_gilgit_road.htm

Ballinger, Pamela, *History in Exile: Memory and Identity at the Borders of the Balkans*, Princeton: Princeton University Press, 2003.

Bender, Barbara and Margot Winer, eds., *Contested Landscapes: Movement, Exile and Place*, Oxford: Berg, 2001.

Boucher, D., *Political Theories of International Relations: From Thucydides to the Present*, Oxford: Oxford University Press, 1998.

Bukhari, Shujaat, 'Apples, Honey Cross LoC', *The Hindu*, 22 October 2008.

———, 'Talks for Opening Kargil, Skardu Road: Manmohan', *The Hindu*, 12 June 2005.

Buzan, Barry, *People, States and Fear: An Agenda for International Security Studies in the Post-Cold War Era*, New York: Harvester Wheatsheaf, 1992.

Cassarino, Jean-Pierre, *Approaching Borders and Frontiers: Notions and Implications*, Florence: Robert Schuman Center for Advanced Studies, European University Institute, 2006.

Chari, P.R. and H.A. Rizvi, 'Making Borders Irrelevant in Kashmir', *Special Report*, United States Institute of Peace, Washington DC, 2008.

Chari, P.R., D. Suba Chandran and S. Akhtar, 'Tourism and Peacebuilding in Jammu and Kashmir', *Special Report*, United States Institute of Peace, Washington DC, 2011.

Dale, Stephen F., 'Silk Road, Cotton Road or . . . Indo-Chinese Trade in Pre-European Times', *Modern Asian Studies*, vol. 43, no. 1, 2009, pp. 79-88.

Delanty, Gerard, 'Irish Political Community in Transition', *The Irish Review*, no. 33, 2005, pp. 13-22.

Doyle, M.W. and G.J. Ikenberry, eds., *New Thinking in International Relations Theory*, Boulder: Westview Press, 1997.

Duffield, Mark, and Nicholas Waddell, 'Securing Humans in a Dangerous World', *International Politics*, no. 43, 2006, 1-23.

Elden, Stuart, 'Contingent Sovereignty, Territorial Integrity and the Sanctity of Borders', *SAIS Review*, vol. 26, issue 1, 2006, pp. 11-24.

Flamenbaum, Stephanie and Megan Neville, 'Optimism and Obstacles in India-Pakistan Peace Talks', *Peace Brief*, USIP, Washington, 2011.

Ganguly, Sumit, *The Crisis in Kashmir: Portents of War, Hopes of Peace*, New Delhi: Foundation Books, 1997.

Grover, Verinder, ed., *The Story of Kashmir: Yesterday and Today*, New Delhi: Deep and Deep Publications, 1995.

Sen Gupta, Babani, *The Fulcrum of Asia*, New York: Pegasus, 1970.

Gupta, Sisir, *Kashmir*, New York: Asian Publishing House, 1966.

Hassnain, F.M., *British Policy Towards Kashmir: 1846-1921*, New Delhi: Sterling Publishers, 1974.

Jha, Prem Shankar, *Kashmir, 1947: Rival Versions of History*, New Delhi: Oxford University Press, 1996.

Johnson, Robert, *A Region in Turmoil*, London and New York: Reaktion, 2005.

Joshi, Nirmala, ed., *Reconnecting India and Central Asia: Emerging Security and Economic Dimensions*, New Delhi: Pentagon Press, 2011.

Kaw, Mushtaq A., 'Restoring India's Silk Route Links with South and Central Asia across Kashmir: Challenges and Opportunities', *The China and Eurasia Forum Quarterly*, vol. 7, no. 2, 2009, pp. 59-74.

Kilot, Nurit and David Newman, eds., *Geopolitics at the End of the Twentieth Century: The Changing World Political Map*, London: Frank Cass, 2000.

Koithara, Verghese, 'Soft Border Approach in Kashmir: How it Can be Made to Work', *Economic and Political Weekly*, vol. 40, no. 18, 30 April-6 May 2005, pp. 1804-6.

Lamb, Alastair, *Kashmir: A Disputed Legacy 1846-1990*, Hertingfordbury: Rexford Books, 1992.

Mahapatra, Debidatta Aurobinda and Seema Shekhawat, 'The Peace Process and Prospects for Economic Reconstruction in Kashmir', *Peace & Conflict Review*, vol. 3, no. 1, 2008, pp. 1-17.

———, *Kashmir Across LOC*, New Delhi: Gyan Publishing House, 2008.

Mahapatra, Debidatta Aurobinda, 'A Perspective on Peace in Kashmir', *ICFAI Journal of Governance and Public Policy,* vol. 2, no. 4, October-December 2007, pp. 31-42.

———, 'Bus Running for Reconciliation?', *Mainstream,* vol. 63, no. 20, 2005, pp. 27-9.

———, 'India's Central Asian Connections', *Journal of Eurasian Studies,* vol. 1, no. 2, 2009, pp. 137-46.

———, 'Mapping Transitional Justice in Kashmir: Drivers, Initiatives and Challenges', *Research Paper,* Oxford Transitional Justice Research, University of Oxford, UK, 2010.

———, 'Poonch for Peace: An Eye Witness Account', *Kashmir Times,* 6 May 2007.

———, 'Silk Route in Kashmir', *Central Eurasian Studies Review,* vol. 8, no. 1, Summer 2009, pp. 13-15.

———, *Central Eurasia: Geopolitics, Compulsions and Connections, Factoring India,* New Delhi: Lancers Books, 2008.

Ministry of Home Affairs, Government of India, *Annual Report, 2010-2011.*

Newman, David, 'On Borders and Power: A Theoretical Framework', *Journal of Borderland Studies*, vol. 18, no 1, 2003, pp. 13-24.

———, 'The Lines That Continue to Separate Us: Borders in Our "Borderless" World', *Progress in Human Geography*, vol. 30, no. 2, 2006, pp. 143-61.

Newman, Edward and Joanne van Selm, eds., *Refugee and Forced Displacement: International Security, Human Vulnerability and the State,* Tokyo: United Nations University Press, 2003.

Newman, Edward, 'Human Security and Constructivism', *International Studies Perspectives*, vol. 2, no. 2, 2001, pp. 239-51.

O'Dowd, Liam and Thomas M. Wilson, eds., *Borders, Nations and States: Frontiers of Sovereignty in the New Europe*, Aldershot: Avebury, 1996.

Raina, N.N., *Kashmir Politics and Imperialist Manoeuvres: 1846-1980*, New Delhi: Patriot Publishers, 1988.

Rudolph, C., 'Sovereignty and Territorial Borders in a Global Age', *International Studies Review*, vol. 7, Issue 1, 2005, pp. 1-20.

Rumford, Chris, 'Theorizing Borders', *European Journal of Social Theory*, vol. 9, no. 2, pp. 155-69.

Schofield, Victoria, *Kashmir in the Crossfire: India, Pakistan and the Unending War*, New Delhi: Viva Books, 2004.

Shekhawat, Seema and Debidatta Aurobinda Mahapatra, *Contested Border and Division of Families in Kashmir: Contextualizing the Ordeal of the Kargil Women*, New Delhi: WISCOMP, 2009.

———, *Kargil Displaced of Akhnoor in Jammu and Kashmir: Enduring Ordeal and Bleak Future*, Geneva: Internal Displacement Monitoring Centre, 2006.

Shekhawat, Seema, *Conflict and Displacement in Jammu and Kashmir: The Gender Dimension*, Jammu: Saksham Books, 2006.

Skjelsbaek, Inger and Dan Smith, *Gender, Peace and Conflict*, New Delhi: Sage Publications, 2001.

Strange, Susan, *The Retreat of the State*, Cambridge: Cambridge University Press, 1996.

'Strengthening Relations across the Line of Control', *Report of the Working Group II, January 2007*, http://www.hinduonnet.com/nic/jk/jkreport_2.pdf.

Waltz, Kenneth, 'Structural Realism after the Cold War', *International Security*, vol. 25, no. 1, 2000, pp. 5-41.

Whitehead, Andrew, *A Mission in Kashmir*, New Delhi: Penguin, 2007.

Wilson, Thomas M. and Hastings Donnan, eds., *Border Identities: Nation and State at International Frontiers*, Cambridge: Cambridge University Press, 1998.

Yaffe, M.D., 'Realism in Retreat? The New World Order and the Return of the Individual to International Relations Studies', *Perspectives on Political Science*, vol. 23, no. 2, 1994, pp. 81-91.

Yusuf, Moeed, *Promoting Cross-LoC Trade in Kashmir: An Analysis of the Joint Chamber*, United States Institute of Peace, Washington DC, 2009.

RCSS PUBLICATIONS

- *Defence, Technology and Cooperative Security in South Asia: Report on the Proceedings of the Tenth Summer Workshop* (RCSS, 2004)
- *Documents on Sri Lanka's Foreign Policy 1947-1965* (RCSS, 2005)
- *Ending the Displacement Cycle: Finding Durable Solutions through Return and Resettlement* (RCSS, 2011)
- *Environment Development and Human Security: Perspectives from South Asia* (2003)
- *Terrorism in South Asia: Impact on Development and Democratic Process* (SIPRA, 2003)
- *South Asia and the War on Terrorism* (India Research Press, 2003)
- *Shaping the Future: A South Asian Civil Society Dialogue* (RCSS, July 2002)
- *Small Arms and Human Insecurity* (RCSS, July 2002)
- *SAARC in the Twenty-First Century: Towards a Cooperative Future* (July 2002)
- *Memories of a Genocidal Partition: The Haunting Tale of Victims, Witnesses and Perpetrators* (RCSS, July 2002)
- *South Asian Security Futures: A Dialogue of Director's Regional Strategic Studies Institutes* (March 2002)
- *Security and Governance in South Asia* (September 2001)
- *Globalization and Non-Traditional Security in South Asia* (September 2001)
- *Security in the New Millennium* (India Research Press, 2001)
- *The Simla Agreement. i972: Its Wasted Promise* (Manohar, 2001)
- *Terrorism in South Asia: Impact on Development and Democratic Process* (RCSS & Konrad Adenauer Foundation, 2003)

RCSS Policy Studies

50 Extremism in Pakistan and India: The Case of Jamaat-e-Islami and Shiv Sena (RCSS, 2010)

49 Democracy as a Conflict Resolution Model for Terrorism: A Case Study of India and Pakistan (RCSS, 2010)

48 Understanding Suicide Terrorism and Bangladesh and Sri Lanka (RCSS, 2010)

47 India's Security Dilemma vis-à-vis China: A Case of Optimum or Sub-Optimum Restraint? (RCSS, 2009)

46 Beyond the Security Impasse: State, Development and People (RCSS, 2008)

45 Conflict Transformation from Ethnic Movement to Terrorist Movement: Case Studies of Tamils in Sri Lanka and Mohajirs in Pakistan (RCSS, 2008)

44 Indo-US Nuclear Cooperation: Altering Strategic Positioning & Shifting Balance of Power in South Asia (RCSS, 2008)

43 Peace Building in Afghanistan: Revisiting the Global War on Terrorism (RCSS, 2008)

42 Getting to Rapprochement over Kashmir: Is Using the 'China Model' a Viable Alternative? (RCSS, 2007)

41 Role of a Third Party in Conflict Resolution: A Case Study of India and Norway in Sri Lanka (RCSS, 2007)

40 International Non-Governmental Organizations in Arms Control and Disarmament Potential and Viability (RCSS, 2007)

39 India-Pakistan Dialogue: Bringing the Society in (RCSS, 2007)

38 The 'People's War' in Nepal (RCSS, 2006)

37 Impact of Partition Refugees in Pakistan: Struggle for Empowerment and State's Response (Manohar, 2006)

36 Maritime Cooperation between India and Sri Lanka (Manohar, 2006)

35 Nuclear Risk Reduction Measures in South Asia: Problems and Prospects (Manohar, 2006)

34 Indo-Pak Conflicts: Ripe to Resolve? (Manohar, 2005)

33 Small Arms and the Security Debate in South Asia (Manohar, 2005)